# Advanced Level
# Practical Physics

The English Language Book Society is funded by the Overseas Development Administration of the British Government. It makes available low-priced, unabridged editions of British publishers' textbooks to students in developing countries. Below is a list of some other books on physics published under the ELBS imprint.

Bennet
**Electricity and Modern Physics**
Edward Arnold

Davies
**Quantum Mechanics**
Routledge

Mandl
**Statistical Physics**
John Wiley

Muncaster
**A-Level Physics**
Stanley Thornes

Noakes
**Structure of Physics**
Macmillan

Pain
**The Physics of Vibrations and Waves**
John Wiley

Rae
**Quantum Mechanics**
Adam Hilger

Welford
**Optics**
Oxford University Press

Wenham *et al.*
**Physics**
Longman

# Advanced Level Practical Physics

by

## M. NELKON,
M.Sc. (Lond.), F.Inst.P., A.K.C.
*Formerly Head of the Science Department, William Ellis School, London*

*and*

## J. M. OGBORN,
M.A. (Cantab.), M.Inst.P.
*Senior Research Fellow, Chelsea College Centre for Science Education, London
and Joint Organiser, Nuffield Advanced Physics Project*

## FOURTH EDITION

HEINEMANN
EDUCATIONAL

Heinemann Educational Books Ltd
Halley Court, Jordan Hill, Oxford OX2 8EJ

Oxford  London  Edinburgh
Madrid  Athens  Bologna  Paris
Melbourne  Sydney  Auckland  Singapore  Tokyo
Ibadan  Nairobi  Harare  Gaborone
Portsmouth NH (USA)

ISBN 0 435 68655 0

Printed and bound in Great Britain by
M & A Thomson Litho Ltd, East Kilbride

# Preface to Fourth Edition

In this edition the experiments have been revised and regrouped to take account of the modernising of the Advanced level syllabus of the major Examining Boards. Among the changes are the inclusion of the following experiments: (1) *Mechanics*—investigation of mechanical resonance, moment of inertia of flywheel, breaking strength and fracture of materials; (2) *Heat*—electrical and mechanical energy methods for specific heat capacity of metal and electrical method for specific latent heat, and temperature measurement with thermistor and thermocouple; (3) *Waves*—transverse and longitudinal waves in springs; (4) *Electricity*—absolute measurement of current, investigation of laws of electromagnetic induction, measurement of flux density in solenoid, measurement of capacitance and of permittivity of air by reed switch, investigation of parallel-plate capacitance by d.c. electrometer, charge and discharge of capacitor; (5) *Electronics and Atomic physics*—$e/m_e$ by using TELTRON tube, junction diode characteristic, $n$–$p$–$n$ transistor characteristics, absorption of beta-particles by aluminium.

Modern apparatus, such as that developed for the Nuffield Advanced level course, has been used where necessary. Experiments have been omitted which are now no longer required for the new syllabuses.

A section of *Short Experiments* has been added. They are intended to illustrate the kind of experimental work required by some Boards, in which part of the examination consists of a number of brief experiments. It is hoped that the selection offered will suggest similar experiments to teachers.

We are indebted to the following for their assistance with this edition: Rev. M. D. Phillips, O.S.B., Ampleforth College, for his valuable contributions and advice in numerous experiments; C. F. Tolman, Whitgift School, Croydon, and M. V. Detheridge, William Ellis School, London, for constructive suggestions throughout the text; P. Betts, Woodhouse School, London, and J. Severn, William Ellis School, London, for considerable assistance in electronic experiments; S. S. Alexander, formerly Woodhouse School, London, for helpful suggestions in many experiments; and T. E. Walton, William Ellis School, London, for his generous aid throughout.

# Preface to First, Second and Third Editions

This book consists mainly of Physics experiments suitable for Advanced level students. They are intended to be a basic or first course at this level, not an exhaustive course of experiments, and no attempt has been made to cater for students proceeding beyond Advanced level.

The book has an introduction on conducting and writing up the experiment, a section on the application of straight-line graphs in measuring physical constants, and a brief treatment of errors and order of accuracy, illustrated by examples. The experiments which follow cover Heat, Optics, Sound, Electricity, and Mechanics and Properties of Matter; the apparatus used is that normally available in schools, and the great majority of experiments can be completed in about an hour and a half. The experiments can be done in any order.

Where possible, we have given brief notes on errors and order of accuracy at the end of the experiments. The errors of measurement depend very much on the apparatus and methods employed, and these will vary widely between schools, but it is hoped that the notes will indicate the lines along which the experimenter should proceed.

A number of miscellaneous examination questions concludes the book, and here the authors are very much indebted to the Senate of London University for permission to reprint questions set in Advanced level practical physics examinations. They are also grateful to M. E. Seymour, B.Sc., formerly of Roan School, Blackheath, for reading the proofs of the book.

In the second edition we added experiments on radioactivity, charge on the electron (Millikan), electron charge–mass ratio, triode valve as amplifier and oscillator, transistor characteristics, cathode-ray oscillograph, a.c. circuits, discharge of capacitor and neon circuit, electric fields, and diffraction grating and photoelectric cell. We are indebted to Mr. T. E. Walton of William Ellis School for his assistance.

The attention of schools is directed to the Department of Education and Science Regulations affecting the use of radioactive materials and high voltages.

For the third edition the text was recast using SI units. Experiments were also added on energy and momentum, microwaves, resolving power of telescope, and radioactive half-life using a d.c. amplifier. In accordance with recent changes in the Advanced Level syllabus, experiments on magnetometry and photometry were omitted. The experiments on the transistor and on radioactivity were regrouped.

# Contents

## PART FIVE. Waves

*Electromagnetic waves—Light*

*Electromagnetic waves—Radio*

*Mechanical waves and Sound*

## PART SIX. Electricity

## PART SEVEN. Atomic Physics including Electronics

## PART EIGHT. Short Experiments

PART ONE

# Introduction

# Conducting the Experiment, and Account

An advanced level practical physics course should enable students to do experiments on the fundamental laws and principles encountered in the theoretical work; to measure a wide variety of physical constants; and to gain experience of a variety of measuring instruments, to learn to handle them with skill, and to appreciate their limitations.

## CONDUCTING THE EXPERIMENT

The experiment should be *planned* in advance; consult reference books to make sure you know the principles underlying the experiment. Before starting the experiment, (i) sketch the circuit if it is an electrical one, or the 'line-up' of the apparatus if it is an optical or complicated experiment, (ii) write down, in the form of a table, all the measurements you propose to make, and the calculation or formula to be used to obtain the result.

After this has been done, set up the apparatus neatly. In electrical circuits, for example, arrange the instruments so that they are well separated, and so that all of them can be reached easily and the scales easily read; in optical experiments always arrange the optical centres of mirrors or lenses to be in line with the object used. Then test the apparatus to see if it is working correctly. A preliminary set of rough readings will show: (i) if the object distances in a lens experiment are too high or too low, for example, or if electrical instruments are reading too high or too low; (ii) the range of values over which measurements may be taken. Enter each result in your table of measurements as it is taken. To obtain accurate results, repeat the experiment using different measurements where possible; for example, vary the object distance in a lens experiment, or the length of wire in a resistivity measurement.

For the same reason, record several values of each measured quantity; for example, take several readings when measuring the diameter of a wire, or the focal length of a converging lens by a plane mirror method. Where possible, as in the measurement of periods of oscillation or of rates of flow of liquids, the measured quantities should be made as large as possible compared with the error in them. For example, in an oscillation experiment the total time measured using a stop-watch should not be less than 100 s, in which case the error in timing should be less than 1%.

The measurements recorded should show the limitations of the instruments used, or the error of observation, so that the final result can be given with a reasonable order of accuracy. This is discussed more fully on p. 7. A graph, plotted roughly while measurements are being made, will often expose a mistake such as the misreading of a meter.

## ACCOUNT OF EXPERIMENT

The account of the experiment should be written in a practical physics notebook, which has graph paper on alternate pages. To be neat and orderly, headings are required at each stage of the account, a suitable scheme being as follows:

1. *Title.*
2. *Diagram.* Use a sharp pencil and rule, and label or letter the parts for reference in your account.
3. *Method* or *Account.* Give full details of the measurements in the order in which they were carried out, mention the precautions you took to obtain accurate measurements, and mention any special methods or techniques used to overcome difficulties or improve the accuracy.
4. *Measurements.* List *all* the measurements taken—do not subtract two measurements and only record the difference, for example. Enter the measurements in a ruled table if possible, or list them below each other in a column, giving the units in which each is measured.

5. *Calculation.* Convert measured values to SI units and substitute them in the formula.
6. *Graph.* If a graph is plotted, choose scales to use as much as possible of the paper, label the two axes, and give the units. Mark the points plotted with crosses, or with a dot and a circle; a sharp, hard pencil is essential.
7. *Result* or *Conclusion.* Your final result must be given to a sensible order of accuracy, and it must state the units concerned; for example, the 'resistivity of the metal was $5.0 \pm 0.2 \times 10^{-7}\,\Omega\,\text{m}$ (p. 10).

For completeness, an outline of the *Theory* of the experiment may be included. *Errors* or uncertainties in the measurements should also be discussed, and the *Order of Accuracy* of the final result then estimated (see p. 7).

## NOTE ON TABLE OF MEASUREMENTS

A *symbol* for a quantity represents both a numerical value and a unit. Thus we may write a length as '$l = 2\,\text{m}$', where *m* represents the unit length, one metre.

The numerical value $2 = l/\text{m}$. Consequently in recording numerical values of *l* in a table of measurements, the heading at the top of the appropriate column would be written '$l/\text{m}$'.

The heading at the top of a column for other quantities would be written in a similar way. Some examples are given below:

| Mass $m/\text{kg}$ | Time $t/\text{s}$ | Temperature $T/\text{K}$ | Current $I/\text{A}$ |
|---|---|---|---|
| | | | |

| Pressure $p/\text{mmHg}$ | Resistance $/\Omega$ | Reciprocal of distance $u^{-1}/\text{cm}^{-1}$ |
|---|---|---|
| | | |

Axes of graphs should be labelled in a similar manner.

# Use of Straight-line Graphs

STRAIGHT-LINE FORMULA

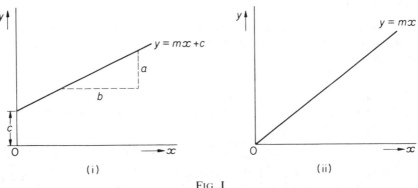

FIG. I

4

If $m$ and $c$ are constants, the general equation of a straight-line graph is

$$y = mx + c$$

(Fig. I(i)). The *gradient* of the line is $m$, since $dy/dx = m$. If a straight line is drawn through points in an experiment, the gradient is measured by finding the ratio $a/b$, where $a$ and $b$ are values read *from the axes of $y$ and $x$ respectively*. Further, when $x = 0$ in the equation $y = mx + c$, then $y = c$. Thus the intercept of the straight line on the $y$-axis is $c$. From this, it follows that the equation $y = 4x - 3$ is a straight line of gradient 4, with an intercept of $-3$ on the $y$-axis. Since $4x - 3 = 0$ when $y = 0$, then $x = \frac{3}{4} =$ intercept on the $x$-axis.

The straight line $y = mx$ differs from $y = mx + c$ in not having the constant $c$. When $x = 0$, then $y = 0$, from $y = mx$; the line therefore passes through the origin (Fig. I(ii)). If $x$ is doubled, then, from $y = mx$, $y$ is doubled; if $x$ is reduced one-third, then $y$ is reduced one-third. In this case, therefore, $y$ is *directly proportional* to $x$. This is not the case with a straight-line equation of the form $y = mx + c$, as the reader should verify.

## APPLICATIONS OF STRAIGHT-LINE GRAPHS

Many quantities in different branches of Physics, such as the resistance of a galvanometer or the focal length of a lens, can be found by plotting a suitable straight-line graph from experimental results. As illustrations of the method, we shall give examples from different branches of the subject.

### *Electricity*

#### 1. *Terminal P.D. Experiment*
The equation for the terminal p.d. $V$ of a cell maintaining a current $I$ in the circuit is

$$V = E - Ir . \quad . \quad . \quad . \quad . \quad . \quad . \quad . \quad . \quad . \quad . \quad . \quad . \quad (1)$$

where $E$ is the e.m.f. and $r$ the internal resistance. This is a straight-line equation between $V$ and $I$ (Fig. II(i)). The gradient of the line, $a/b$, is numerically $r$, the coefficient of $I$, and this can therefore be found. Further, when $I = 0$, $V = E$; thus the e.m.f. is the intercept on the $V$-axis, as shown.

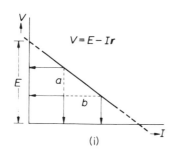
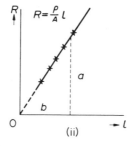

FIG. II

#### 2. *Resistivity Experiment*
With the usual notation, the resistance $R$ of a wire is given by

$$R = \rho \frac{l}{A} \quad . \quad . \quad . \quad . \quad . \quad . \quad . \quad . \quad . \quad . \quad . \quad . \quad (2)$$

Thus, if the resistances $R$ of several lengths $l$ are measured and plotted against $l$, a straight-line graph is obtained which passes through the origin (Fig. II(ii)). The gradient, $a/b$, of the line is the coefficient of $l$ in the equation. Hence $\rho/A = a/b$, or $\rho = A \times (a/b) = \pi r^2 \times (a/b)$, where $r$ is the radius of the wire. Thus by measuring the gradient of the line, and knowing the radius, the resistivity $\rho$ can be calculated.

### 3. *Diode Valve Experiment*

In a diode valve the anode current $I_a$ may be related to the anode voltage $V_a$ by the formula $I_a = kV_a^n$, where $k$ and $n$ are constants. To find $n$, we take logs of both sides. Then

$$\log I_a = \log k + n \log V_a. \quad \ldots \ldots \ldots \ldots \quad (4)$$

Thus a graph of $\log I_a$ v. $\log V_a$ is a straight line, and the coefficient $n$ of $\log V_a$ is the gradient of the line. Generally, it should be remembered that an unknown power or index can be found by plotting logarithms of the quantities concerned.

## Optics

### 4. *Newton's Rings Experiment*

In this experiment, the radius $r_n$ of the $n$th ring is related to the wavelength $\lambda$ of the incident light and the radius $R$ of the lower lens surface by the relation

$$r_n^2 = (n + \tfrac{1}{2})R\lambda = nR\lambda + \tfrac{1}{2}R\lambda. \quad \ldots \ldots \ldots \ldots \quad (5)$$

So a graph of $r_n^2$ against $n$ is a straight line whose gradient is $R\lambda$. Hence $\lambda = gradient/R$ and thus $\lambda$ can be calculated.

### 5. *Displacement Experiment*

In the displacement method of measuring the focal length of a converging lens,

$$f = \frac{l^2 - d^2}{4l}. \quad \ldots \ldots \ldots \ldots \ldots \quad (6)$$

where $l$ is the distance between the object and screen, and $d$ is the lens displacement.

$$\therefore \quad \frac{l}{4} - \frac{d^2}{4l} = f$$

or

$$l = \frac{d^2}{l} + 4f$$

Thus, when a graph of $l$ v. $d^2/l$ is plotted, a straight-line graph is obtained. Further, since $l = 4f$ when $d^2/l = 0$, the focal length can be found by producing the graph back to cut the $l$-axis.

## Sound

### 6. *Resonance Tube Experiment*

For the first resonance position, we have

$$\frac{\lambda}{4} = l + c = \frac{V}{4f} \quad \ldots \ldots \ldots \ldots \ldots \quad (7)$$

with the usual notation. Thus, when $f$ is varied and $l$ is measured, a graph of $l$ v. $1/f$ is a straight line. The gradient of the line is $V/4$, since this is the coefficient $(m)$ of $1/f$, and hence $V$ can be found. Also, when $1/f = 0$, $l + c = 0$ or $l = -c$. Thus, the negative intercept of the graph on the $l$-axis is the end-correction $c$.

### 7. *Sonometer Experiment*

If an unknown weight $X$ is placed on a sonometer wire, and additional known weights $W$ are added, then, with the weights in newtons,

$$f = \frac{1}{2l}\sqrt{\frac{X + W}{m}} \quad \ldots \ldots \ldots \ldots \ldots \quad (8)$$

Squaring and cross-multiplying, $4mf^2l^2 = X + W$.

Thus, if $W$ is varied, and $l$ is altered each time until the wire is tuned to the same frequency $f$, a graph of $l^2$ v. $W$ is a straight line. Now when $l^2 = 0$, $X + W = 0$, or $W = -X$. The unknown weight $X$ is therefore the negative intercept on the $W$-axis of the straight-line graph.

# Errors, and Order of Accuracy

The errors in a particular experiment may be due to the observer, or to the instrument used, or to a combination of both. They may also be present by the very nature of the experiment; for example, the heat lost by cooling in a heat experiment causes an error in the final temperature obtained. In the latter cases special techniques can minimise or eliminate this error, for example by making a cooling correction, or by repeating the experiment and subtracting to eliminate the heat lost. However, when errors or uncertainties are present in experiments we must record them in the measurements, and take them into account when calculating the final result, as discussed later.

## ERRORS OF OBSERVATION OF INSTRUMENTS

If a liquid temperature is read as $15.2\,°C$ on a thermometer calibrated in degrees, the temperature should be recorded as $15.2\pm0.1\,°C$ if readings are estimated to the nearest 1/10th °C, to denote the possible temperature range $15.1–15.3\,°C$. When the liquid is warmed to an observed temperature of $39.8\,°C$ this should be recorded as $39.8\pm0.1\,°C$. The temperature increase, by subtraction, is $24.6\pm0.2\,°C$, as the possible temperature rise may vary from 24.4 to 24.8 °C. Similarly, if a calorimeter weighs $44.3\pm0.1\,g$, and the mass of the calorimeter with some water is $84.9\pm0.1\,g$, then the mass of water is $40.6\pm0.2\,g$. Generally, then, errors are *added* when two similar quantities are subtracted; the errors are also added, of course, when two similar quantities are added.

It is important to realise that an error may occur at the 'zero' of a measuring instrument, as well as at the actual reading on the instrument. For example, suppose a stop-watch is graduated in intervals of $\frac{1}{5}$ s, and the time for a number of oscillations of a pendulum is observed. An error of $0.2$ s may then occur on starting the watch at the centre of an oscillation; thus if the time on the watch after the interval concerned is $54.6$ s, the time should be recorded as $54.6\pm0.4$ s, which takes into account the errors at the beginning and end. Similarly, if a ruler is graduated in millimetres a length may be recorded as $62.2\pm0.2$ cm to take into account errors at both ends of the length. When in doubt, always over-estimate, rather than under-estimate, the error.

### Setting Errors

In some experiments uncertainties or errors may occur when a setting is made. In a lens experiment, for example, there is an uncertainty in deciding when the image of crosswires on a screen is in sharpest focus. The uncertainty or error here can be found by moving the screen until the image is *just* blurred and noting the distance from this position to that when the image was in sharpest focus. The metre bridge and potentiometer experiments depend on finding a balance point; the uncertainty or error here may be found by moving the jockey or slider until the galvanometer pointer is just deflected and noting the distance from this position to the 'best' position of balance. In weighing, the uncertainty can be obtained by finding the smallest additional weight which will just move the balance pointer. These examples show that uncertainties of this type can be estimated by altering the quantity concerned until the difference is just perceptible.

## PERCENTAGE ERROR

When a temperature is recorded as $15.2\pm0.1\,°C$, the percentage error is $(0.1/15.2)\times100$ or about $0.7\%$. With a temperature of $35.2\pm0.2\,°C$ the percentage error is $(0.2/35.2)\times100$, or about $0.6\%$. When the temperatures are added the result is $50.4\pm0.3\,°C$, which is an error of about $0.6\%$. If the temperatures are subtracted, however, the result is $20.0\pm0.3\,°C$, which is an error of about $1.5\%$.

Suppose that the radius $r$ of a wire is measured as $0.56\pm0.02$ mm, a percentage error of about $3.5\%$. If the percentage error in $r^2$ is needed because the area, $\pi r^2$, enters into a formula, for example in measuring resistivity, then

$$\text{Error in } r^2 = (0.56 \pm 0.02)^2 - 0.56^2 = \pm 2 \times 0.56 \times 0.02, \text{ approx.}$$

and

$$\text{percentage error in } r^2 = \frac{2 \times 0.56 \times 0.02}{0.56^2} \times 100$$

$$= \frac{2 \times 0.02}{0.56} \times 100 = 7\% \text{ approx.}$$

The percentage error in $r^2$ is thus *twice* the percentage error in $r$. To show this is true for small percentage errors,

$$\text{error in } r^2 = \delta(r^2) = 2r \cdot \delta r$$

$$\therefore \text{ percentage error in } r^2 = \frac{\delta(r^2)}{r^2} \times 100 = \frac{2r \cdot \delta r}{r^2} \times 100 = 2\frac{\delta r}{r} \times 100 = 2 \times \text{percentage error in } r$$

Similarly,

$$\text{percentage error in } r^4 = \frac{\delta(r^4)}{r^4} \times 100 = \frac{4r^3 \cdot \delta r}{r^4} \times 100 = 4\frac{\delta r}{r} \times 100$$

$$= 4 \times \text{percentage error in } r$$

If powers of a quantity $X$ occur in an experiment, then, the percentage error will be large unless $X$ itself is measured to a high degree of accuracy, a point which should be remembered.

### Percentage Errors in Products and Quotients

In many cases in Physics, we set out to measure a quantity which depends on the product or quotient of observable quantities. Thus in measuring the resistivity, $\rho$, of a material (see p. 140) the formula

$$\rho = \frac{R \cdot \pi d^2}{4l} \quad \ldots \ldots \ldots \ldots \quad \text{(i)}$$

is used. If errors in $R$, $d$, and $l$ are made, and represented by $\delta R$, $\delta d$, and $\delta l$ respectively, the percentage error in $\rho$ can best be found by taking logs of both sides in the formula for $\rho$, and then differentiating. Thus, from (i),

$$\log \rho = \log R + \log \pi + 2 \log d - \log 4 - \log l \quad \ldots \ldots \ldots \quad \text{(ii)}$$

and by differentiation,

$$\frac{\delta \rho}{\rho} = \frac{\delta R}{R} + \frac{2\delta d}{d} - \frac{\delta l}{l} \quad \ldots \ldots \ldots \ldots \quad \text{(iii)}$$

Each term on the right side, when multiplied by 100, represents the percentage error in $R$, $d$, and $l$, and hence the percentage error in $\rho$ can easily be found. Since the percentage errors can be positive or negative, it follows that

$$\frac{\delta \rho}{\rho} = \pm \frac{\delta R}{R} \pm \frac{2\delta d}{d} \mp \frac{\delta l}{l}$$

and thus the maximum percentage error is obtained by *adding* the percentage error in $l$ to those in $R$ and $d$.


ORDER OF ACCURACY

It has sometimes been said that the final result of an experiment has little meaning unless the limits of error or the order of accuracy is also stated. Suppose the sum of the percentage errors in a focal length experiment is 2%, for example, and the focal length is found by logarithms to be 26·47 cm. It is then incorrect to give the final result as $26.47 \pm 0.01$ cm, because this implies an error of 1 part in 2647, which is about 0·04%. If the result is given as $26.5 \pm 0.1$ cm the percentage error is 1 part in 265, or about 0·4%, which is still incorrect. It can thus be seen that the final result should be given as $26.5 \pm 0.5$ cm, in keeping with the error of 2% in the experiment.

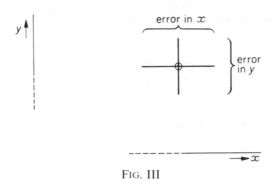

Fig. III

## GRAPHS

If graphs are plotted, each point may be represented by a cross drawn with horizontal and vertical lines or arms, the lengths of the arms representing respectively the errors in each of the two readings $x$ and $y$ (Fig. III). The 'best' straight line is the one which passes through both arms of each cross as near as possible to the point of intersection. The straight line is best found with a thread stretched across the paper, or with a transparent ruler. The uncertainty or error in the slope may be found by drawing the line of greatest or least slope which still just cuts the two lines of the vertical crosses, and finding the difference $D$ between this slope and that $A$ of the 'best' straight line. This is the maximum error in the slope. If the intercept of the line on an axis is required, a similar procedure will enable the maximum error $D$ in the intercept $A$ to be found. In either case the approximate value of the order of accuracy is given by $(D/A) \times 100\%$.

## AVERAGES

If many experiments are carried out to determine the value of a quantity, the average, $A$, gives a better estimate of the true result than any of the results singly. If the difference between the average, $A$, and the result differing most greatly from it is $D$, an approximate value of the order of accuracy may be obtained by calculating $(D/A) \times 100\%$.

## FORMULAE

Suppose the measurements recorded in determining the resistivity $\rho$ of a metal are:

Resistance of wire $= 2.06 \pm 0.01\, \Omega\ (\pm 0.5\%)$
Diameter of wire $= 0.57 \pm 0.01$ mm $= (0.57 \pm 0.01) \times 10^{-3}$ m $(\pm 1.8\%)$
Length of wire $= 105.6 \pm 0.1$ cm $= (105.6 \pm 0.1) \times 10^{-2}$ m $(\pm 0.1\%)$

*Calculation omitting errors.*
$$\rho = \frac{R . \pi d^2}{4l} = \frac{2.06 \times \pi \times (0.57)^2 \times 10^{-6}}{4 \times 105.6 \times 10^{-2}}$$
$$= 4.977 \times 10^{-7}\, \Omega\, \text{m}$$

*Percentage error.* From the formula for $\rho$,

$$\frac{\delta\rho}{\rho} = \pm\frac{\delta R}{R} \mp \frac{2\delta d}{d} \pm \frac{\delta l}{l}$$

$\therefore$ maximum percentage error $= +0.5 + 2 \times 1.8 + 0.1 = 4.2\%$

*Order of accuracy.* The error may thus be 4·2% of $4.977 \times 10^{-7}$. Clearly it is incorrect to give the result for $\rho$ to four or to three significant figures, since this implies a percentage error of 1 in 5000 or 1 in 500 respectively, which is 0·02% or 0·2%. As the error is of the order of 4%, the resistivity

9

result should be given to two significant figures, as

$$\rho = 5 \cdot 0 \pm 0 \cdot 2 \times 10^{-7} \, \Omega \, \text{m}$$

*Note.* As the error in the diameter predominates, the diameter of the wire must be measured with even greater accuracy to increase the accuracy of the final result.

It should be carefully noted that some measurements are inevitably made with a relatively large degree of uncertainty or error; for example, a thermometer reading in tenths of a °C is usually the most accurate obtainable in a school laboratory, and hence the temperature rise in the electrical experiment to measure specific heat capacity of a metal may be $5 \cdot 8 \pm 0 \cdot 2$ °C, a percentage error of $3 \cdot 5\%$. In this experiment, a little thought shows it is unnecessary to weigh the metal to a very high degree of accuracy, such as $501 \cdot 3$ g for example, because this is a percentage error of $(1/5013) \times 100$ or $0 \cdot 02\%$, which has negligible effect on the total error. Measuring the mass to the nearest 1 g, 501 g, would be sufficient; this is a percentage error of about $0 \cdot 2\%$.

PART TWO

# Mechanics and Properties of Matter

# EXPERIMENT 1

## Measurement of g by Simple Pendulum

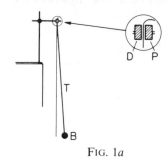

FIG. 1a

### APPARATUS

Spherical bob B, thread T, stop-watch, tall stand and clamp, cork pads D, P, pointer for reference mark.

### METHOD

Attach a long piece of thread, about 2 metres long, to the pendulum bob. Fix the top of the thread between the cork pads D, P, placed in the jaws of the clamp, so that the bob just clears the floor. Place a reference mark, using a pointer, at the equilibrium position of the bob. Now set the bob oscillating through a *small* angle, and beginning your counting through the equilibrium position of the bob, find the time for $N$ (e.g. 30) complete oscillations. Measure the length $l$ of the thread from the point of suspension to the point of attachment of the bob, and enter the result in metres (m).

By raising the thread each time, diminish the length of the pendulum by about 25 cm, 50 cm, 75 cm, and 100 cm. On each occasion find the length of the thread in metres and the time for $N$ complete oscillations.

### MEASUREMENTS

| Length $l$/m | Time for $N$ oscillations /s | Period $T$/s | $T^2$/s$^2$ |
|---|---|---|---|
|  |  |  |  |

### CALCULATION

Find the period, $T$, the time for one complete oscillation, then calculate $T^2$ and enter in the table of measurements.

## GRAPH

Plot $l$ v. $T^2$, and draw the best straight line passing through the points. See graph, Fig. 1$b$. Measure the gradient, $a/b$ of the line.

Since
$$T = 2\pi\sqrt{\frac{l+c}{g}}$$

where $c$ is the small constant length from the point of attachment of the bob to the centre of gravity of the bob,

then, simplifying,
$$l = \frac{g}{4\pi^2}T^2 - c$$

$$\therefore \ g = 4\pi^2 \times \text{gradient of line} = \ \ldots$$

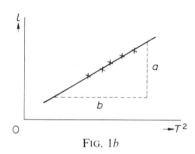

FIG. 1$b$

## CONCLUSION

The acceleration due to gravity was .. $\mathrm{m\,s}^{-2}$.

## ERRORS

The error in timing may be estimated by noting the interval (e.g. 1/5 s) at which the hand moves. Errors in timing occur when the watch is started and when it is stopped. The number of swings should be large enough for the total error to be less than 1% of the time recorded. If the clamp is not rigid, the effective length of the pendulum is increased, a systematic error.

## ORDER OF ACCURACY

Find the change in the slope of the line caused by drawing lines of greatest and least slope which just agree with the plotted points. The error in $g$ is then $4\pi^2 \times$ the error in the slope.

# EXPERIMENT 2

# Measurement of Elastic Constant of Spiral Spring, and Earth's Gravitational Intensity

## ELASTIC CONSTANT OF SPRING

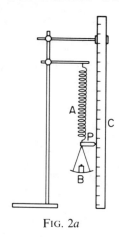

FIG. 2a

APPARATUS

Light spiral spring A, scale-pan B, metre rule C, two clamps and stands, two boxes of weights, light pointer P for spring, stop-watch.

METHOD

Suspend the light spring from the clamp, and attach a light pointer P to the spring. Set up a fixed vertical metre rule C beside the spring A, attach B to the spring, and then add suitable weights, noting the reading of the pointer each time. Do this for about eight loads on the scale-pan. Then remove each weight, and record the reading of P each time. If the spring has not been permanently strained the reading of P will return to its original or 'zero' reading when all the weights and the scale-pan have been removed. Weigh the scale-pan.

MEASUREMENTS

Mass of scale-pan = .. kg
Zero reading of spring = .. mm

| Mass on scale-pan /kg | Reading on metre rule /mm | Total mass /kg | Extension /mm |
|---|---|---|---|
|  |  |  |  |

14

## CALCULATION

Add the scale-pan mass to each load to find the total mass and enter the results in the table of measurements in kilogram.

## GRAPH

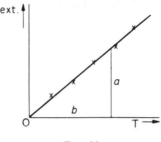

FIG. 2*b*

Plot the extension in metre v. the mass T whose weight extended the spring (Fig. 2*b*). Draw the best straight line through the origin. From the graph, calculate $\mu$, the mass in kg per metre extension; this is $b/a$ (see Fig. 2*b*).

## CONCLUSION

1. $\mu=$ .. kg per metre extension is the mass hung on the spring per metre extension.
2. If the graph is a straight line passing through the origin, then the extension of the spring is directly proportional to the tension in the spring (Hooke's law).

## ERRORS

Errors occur in reading the pointer positions and in weighing.

## ORDER OF ACCURACY

From the origin draw (i) the line of least slope through the points; (ii) the line of greatest slope through the points. Find the error in $b/a$ from the changes in the slope.

*(Experiment continued overleaf)*

# EARTH'S GRAVITATIONAL INTENSITY

METHOD

Using the spring and scale-pan as before, add a suitable mass to the scale-pan. Pull the scale-pan down slightly and release it, so that the spring makes simple harmonic oscillations. Time $N$ (e.g. 30) complete oscillations. Add increasing masses to the scale-pan, observe the time for $N$ complete oscillations on each occasion, taking five sets of loads and times.

MEASUREMENTS

Mass of scale-pan = .. kg

| Load on scale-pan /kg | Time for $N$ oscillations /s | Load on spring $M$ /kg | Period $T$ /s | $T^2$ /s$^2$ |
|---|---|---|---|---|
| | | | | |

CALCULATION

Add the scale-pan mass to the load to find the total load $M$ on the spring. Calculate the period $T$, and then $T^2$, and enter in the table of measurements.

GRAPH

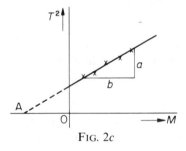

FIG. 2c

Plot $T^2$ v. $M$, the mass on the spring in kilogram (kg) (Fig. 2c). Draw the best straight line through the points. Measure the gradient, $a/b$, of the line.

## THEORY

The period $T$ is given by

$$T = 2\pi \sqrt{\frac{M+m}{\lambda}}$$

where $m$ is a constant depending on the mass of the spring itself and $\lambda$ is the elastic constant of the spring in newton per metre.

$$\therefore \quad \frac{\lambda}{4\pi^2} T^2 = M + m \quad . \quad . \quad . \quad . \quad . \quad . \quad . \quad . \quad . \quad . \quad . \quad (1)$$

$$\therefore \quad \text{gradient of } T^2 \text{ v. } M \text{ graph is } \frac{4\pi^2}{\lambda}$$

$$\therefore \quad \lambda = 4\pi^2 / \text{gradient } a/b \quad . \quad . \quad . \quad . \quad . \quad . \quad . \quad . \quad . \quad (2)$$

Thus a force $\lambda$ per metre extends the spring. From p. 16, the earth exerts a gravitational force $\lambda$ on a mass $\mu$.

$$\therefore \quad \text{Earth's gravitational intensity } g = \frac{\lambda}{\mu} = \quad .. \quad N\,kg^{-1}$$

## CONCLUSION

The Earth's gravitational intensity $g = \quad .. \quad N\,kg^{-1}$

## ERRORS

Errors occur in the measurement of the period and in the scale-pan mass.

## ORDER OF ACCURACY

From above, maximum percentage error in $g$

$$= \left[ \frac{\delta\lambda}{\lambda} + \frac{\delta\mu}{\mu} \right] \times 100\%$$

The percentage error in $\lambda$ is obtained from the first part of the experiment; the percentage error in the gradient can be found by drawing the lines of greatest and least slope which pass through the majority of the points and comparing it with the slope of the 'best' line.

# EXPERIMENT 3

# Investigation of Mechanical Resonance

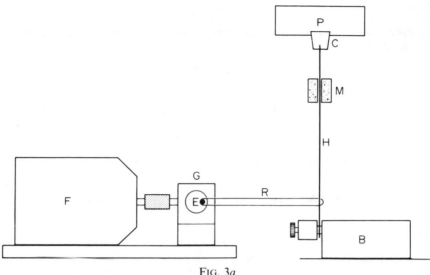

FIG. 3a

APPARATUS

Hacksaw blade oscillator (e.g. Griffin & George, L61–490)—consisting of hacksaw blade H, base B to hold the blade firmly in a vertical position, and sliding mass M, slotted cork C and piece of card (e.g. postcard) P; rubber band R, fractional horsepower motor, 12 V, F (e.g. Griffin & George, N14–540); reduction gearbox G, eccentric drive E; low-voltage d.c. variable power supply 0–25 V; G-clamps; stop-watch; metre rule.

METHOD

*Preliminary.* Clamp the hacksaw blade H to its base B, and tighten the nuts holding it in place until it is firmly held in a vertical position. Slip a rubber band R over the blade, and loop the band round an eccentric (off-centre) peg or screw E let into the pulley of the gearbox G. The gearbox is driven by the fractional horsepower motor F. It should turn at between one revolution in about 5 seconds to about five revolutions in 1 second, over the range of motor speeds provided by varying the voltage of its 0–25 V d.c. power supply.

(*Alternatively*, the motor and drive can be replaced by a massive pendulum of adjustable length, which is kept swinging by gently pushing it with a finger.)

Cut a cross on the ends of the cork C and fit the cork onto the top of the blade H as shown in Fig. 3a. Fit a card P into the cork so that the card oscillates in its own plane, that is, it moves edge on.

Slip the rubber band R temporarily off the eccentric drive E and move the mass M up or down the blade until the natural frequency of oscillation of the blade, produced by a slight displacement, is about 2 or 3 oscillations per second.

Put the rubber band back on the eccentric drive, position the apparatus so that the band is always in tension when the drive rotates, and clamp the whole apparatus to the bench.

18

*Experiment.* 1. Set the drive rotating at about one revolution a second, and observe the motion of the blade carefully. It is being driven at a frequency less than its natural frequency, and for a time it will keep starting and stopping oscillating (transient oscillations). Wait until it is oscillating steadily at the driving frequency, put a ruler behind the top of the blade, and record the amplitude of the oscillations (twice the amplitude—the distance between extreme positions—will do as well). Use the stop-watch to find the time taken for say 20 oscillations, and record the time.

2. Increase the speed of the motor, by raising the voltage supplied to it, and repeat these two observations for at least five, preferably ten, different driving frequencies. (Note that when the driving frequency is close to the natural frequency of the blade, the transient oscillations rise and fall very slowly, so that a transient oscillation can be mistaken for a steady oscillation.) Make sure you have about as many observations at frequencies above the natural frequency as below.

3. Now make the air damping of the blade heavier by turning the card through a right angle, so that it moves at right angles to its plane (that is, moving face on). Repeat the previous observations.

MEASUREMENTS

| | Number of oscillations, $N$ | Time for oscillations $t$/s | frequency $f$/Hz | amplitude $a$/mm |
|---|---|---|---|---|
| Light damping | | | | |
| Heavy damping | | | | |

CALCULATION AND GRAPH

Calculate the frequency $f = N/t$ for each observation, and enter values in the table of results.

Plot a graph of amplitude $a$ against frequency $f$ for light and heavy damping, drawing a smooth curve passing close to the plotted points. Fig. 3$b$. If necessary, make more measurements near the maximum of the curve, where the graph is hardest to draw.

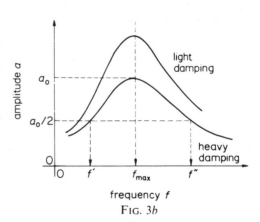

FIG. 3$b$

CONCLUSIONS

Describe the shape of the curve obtained for each case, and the differences between the two curves. Look particularly at
(i) the frequency at which the amplitude is a maximum,
(ii) whether the curve looks symmetrical about the maximum or not,
(iii) whether the 'width' of the curve changes with the amount of damping—the 'width' of the curve can be estimated as the ratio $(f'' - f')/f_{max}$, where $f_{max}$ is the frequency for maximum amplitude and $f'$ and $f''$ are the frequencies either side of the maximum at which the amplitude is half the maximum value.

# EXPERIMENT 4

# Measurement of Moment of Inertia of Flywheel

## APPARATUS

Flywheel (usually mounted for rotation about a horizontal axis), thin string or cord, suitable weight (e.g. 0·5 kg), metre rule, calipers, stop-watch.

## METHOD

1. Attach one end of the string to the mass W on the ground and cut off a length of string whose free end passes *once* round the axle with W on the ground.

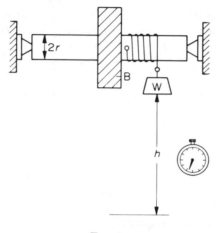

FIG. 4

2. Now wind the string evenly round the axle until the mass W rises to its highest point as shown— pass the string through the hole in the axle if present and overlap the first few turns, so that the string does not slip when it unwinds. Check that, when the weight just reaches the ground, the string completely leaves the axle—if not, readjust the length of string tied to W. If necessary, oil the bearings of the axle of the flywheel.

3. Measure the height $h$ of the mass W above the ground. Make a chalk mark B on the flywheel. Now allow W to fall to the ground and make the following three measurements: (i) the time $t$ taken for W to reach the ground, (ii) using B, the number of revolutions $n_1$ of the wheel in this time, (iii) the number of revolutions $n_2$ of the wheel before coming to rest *after* W has reached the ground. Check the value of $n_1$ from the relation $n_1 = h/2\pi r$, where $r$ is the radius of the axle. Using the calipers, take several readings of the axle diameter, $2r$.

4. Rewind the string round the axle and obtain two more readings of $h$, $t$, $n_1$ and $n_2$, starting from the same position of the mass above the ground each time. The results should be fairly consistent— if not, the frictional couple in the bearings has altered and more lubrication is required.

## MEASUREMENTS

Mass $m$ of W = .. kg          Height $h$ = .. m
Axle diameter, $2r$ = .., .., .. m, average = .. m

|                | (1) | (2) | (3) |
|----------------|-----|-----|-----|
| Time $t/s$     |     |     |     |
| No. of revs $n_1$ |     |     |     |
| No. of revs $n_2$ |     |     |     |

## CALCULATION

From the energy equation for the motion of the flywheel,

$$mgh = \tfrac{1}{2}I\omega^2 + \tfrac{1}{2}mr^2\omega^2 + n_1 f \quad \ldots \ldots \ldots \ldots \quad (1)$$

$$n_2 f = \tfrac{1}{2}I\omega^2 \qquad\qquad \ldots \ldots \ldots \ldots \quad (2)$$

where $f$ is the work done against friction per revolution. Eliminating $f$ from (1) and (2), and using $\omega = 2h/tr$, we obtain finally

$$I = mr^2 \left( \frac{gt^2}{2h} - 1 \right) \left( \frac{n_2}{n_1 + n_2} \right)$$

Substitute the mean values for $m$, $h$, $r$, $t$, $n_1$, $n_2$ and $g = 9 \cdot 8 \, \mathrm{m\,s^{-2}}$ and calculate $I$.

## CONCLUSION

The moment of inertia of the flywheel was found to be .. kg m².

## ERRORS

Discuss the effect on your result of any change in the frictional force in the bearings while the experiment was carried out, and of any measurement difficult to make accurately. Estimate the order of accuracy from your different results, which should then be reflected in your stated result for the moment of inertia.

# EXPERIMENT 5

# Kinetic Energy and Linear Momentum—
# Investigations with Air-track

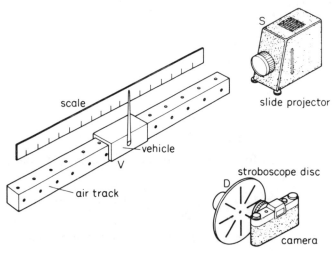

FIG. 5a

## APPARATUS

Air-track; vehicles V; air blower (vacuum cleaner); stroboscope disc D and motor; 35 mm camera and stand; film: KODAK one-bath developer; bucket of water; 35 mm cardboard mounts; slide projector S.

## INTRODUCTION

*Air-track.* An 'air-track' is a device for obtaining motion in which there is very little friction (Fig. 5a). The vehicle V rides on a thin cushion of air, which is blown from inside the track through many small holes in its surface. The apparatus makes a good 'spirit level'—a slight tilt sends the vehicle accelerating down the slope. The track needs careful levelling, until a vehicle will move slowly either way at a steady speed.

*Photographic method for velocity.* In these experiments, the velocity of vehicles on the track will require measurement. This may be done with the aid of a lamp and a photodiode, which operate a scaler or fast clock when the vehicle obscures the light shining on the diode. The use of a photographic method, however, is suggested, with a scale for distance travelled included in the picture as shown in Fig. 5a.

The camera is fixed perhaps 2 m from the track and is focused on the track. A rotating slotted disc D is then placed in front of the camera lens. With the shutter speed dial set to 'B', the shutter opens when the trigger is pressed and closes only when it is released. While the shutter is open, the rotating slots in D provide the camera with a series of equally-timed glimpses of the moving vehicle V. A white marker on V, such as a length of milk straw, gives a good image. From the distances moved in these times, as recorded on the film, the velocity of V can be found. The vehicle is best illuminated by a beam of light from a slide projector S, which shines along the track.

*Developing.* The KODAK leaflet 'Record Photography in Schools' describes a simple way for developing film without need for a dark room. 40 cm$^3$ of combined developer and fixer are needed in a small glass or beaker, and the exposed film inside the cassette is placed in the solution. Film lengths of up to 20 exposures are suitable. Before putting it into the solution, the film must be wound tightly into the cassette, holding down the end of the film on the outside with a rubber band. Once in, the film is unwound a few turns, when air bubbles out of the cassette. The core of the cassette is then turned backwards and forwards by one or two turns each way for the four minutes of development and fixing time. The cassette can then be dropped into a bucket of water, and the developed film pulled out and cut off. Individual frames can be mounted in cardboard mounts and projected onto a screen or sheet of paper for viewing.

In some cameras, the 'take-up' spool can be replaced by an empty cassette. If so, less film need be wasted when only a few exposures are required. In this case the free end of the film is taped to the core of the empty cassette, the cassettes are replaced in the camera, and after the film is exposed, this part of the film is wound onto the take-up cassette. The camera is then opened, the film between the cassettes is cut and the film in the take-up cassette is developed.

## KINETIC ENERGY AND SPEED

1. A vehicle can be given kinetic energy by catapulting it down the air-track using an elastic band. Most air-tracks provide a pair of catapult arms between which elastic may be stretched Fig. 5b). If the vehicle is pulled back a fixed distance (perhaps 5 cm) and released, it will be given

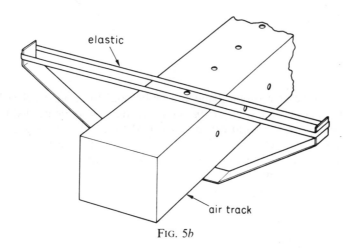

elastic

air track

FIG. 5b

some energy, $E$. If it is pulled back twice as far and then released, it will *not* be given twice the energy, for several reasons. Can you think of one or more? But two, three and four exactly similar rubber bands can be used to give energies $2E$, $3E$, $4E$ to the vehicle. How can this be done? How would you test that the bands were similar?

2. Take a number of measurements in which the vehicle velocity $v$ is found for energies of $E$, $2E$, $3E$ and $4E$ respectively. Enter the results in a table:

| Energy | $E$ | $2E$ | $3E$ | $4E$ |
|--------|-----|------|------|------|
| $v$ |  |  |  |  |

*Plot a suitable graph* to test whether the kinetic energy is proportional to $v^2$. (There is more than one way of doing this. Can you think of one which differs from the graph you used?)

*(Experiment continued overleaf)*

## 5. (*Continued*)

3. Vehicles of different mass can be used to investigate the variation of kinetic energy, K.E., with mass. For a constant K.E., how would the mass $m$ and velocity $v$ be related? Make a series of measurements of $m$ and $v$ to test this relationship and then plot a suitable graph.

### ENERGY STORED IN STRETCHED ELASTIC AND SPRING

If the force $F$ needed to stretch a spring or length of elastic is directly proportional to the extension $x$, the energy stored at this extension is $\frac{1}{2}F.x$ (show this is the case) (Fig. 5c (i)). For elastic, this behaviour, in which Hooke's law is obeyed, is not typical. The $F-x$ graph looks more like Fig. 5c (ii) than Fig. 5c (i). The energy stored is then *not* $\frac{1}{2}F.x$. Is it less or more?

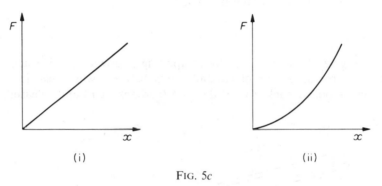

(i)          (ii)

FIG. 5c

*Elastic.* A length of elastic or a rubber band can be hooked to an air-track vehicle V so that the elastic is released when it goes slack (Fig. 5d (i)). (Some ingenuity may be needed.) The force $F$ needed to produce the extension $x$ in the band can be measured by means of weights hanging at the end of a string over a pulley. Alternatively, a spring balance can be attached to the end of the band. After the vehicle, mass $m$, is released its velocity $v$ can be found from the photograph taken and the kinetic energy, $\frac{1}{2}mv^2$, can be calculated. This can be compared with the energy $\frac{1}{2}F.x$ stored in the band, assuming it obeys Hooke's law. How do you expect the measured energies to compare? How far is this borne out by your results?

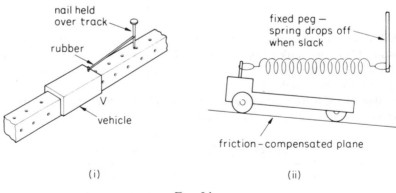

(i)          (ii)

FIG. 5d

*Steel spring.* It is very hard to catapult an air-track vehicle with a spring. Also, too much energy may be left behind as motion of the spring. However, it is quite easy to catapult a dynamics trolley (about 1 kg) down a friction-compensated slope. This is a slope tilted slightly until the trolley runs

24

down at a *steady* velocity, when the frictional forces are balanced by the component of the weight along the slope. Fig. 5d (*ii*) shows one arrangement—you may think of a better one. If the extension $x$ is varied, you should find $v \propto x$. Why?

LINEAR MOMENTUM

If two vehicles of masses $m_1$, $m_2$ and velocities $v_1$, $v_2$ respectively collide head-on, then, from the principle of the conservation of linear momentum,

$$(m_1v_1 + m_2v_2)_{\text{BEFORE}} = (m_1v_1 + m_2v_2)_{\text{AFTER}}$$

It must be remembered that momentum is a vector quantity. Thus a pair of vehicles moving in opposite directions, both with mass $m$ and velocity $v$, each have a momentum $mv$, but their *total* momentum $= mv - mv = 0$.

By photographing a series of collisions of different kinds, the principle of the conservation of linear momentum can be tested. Some suggestions, using two vehicles of equal or different mass, are:

  (i) one still, the other moving and sticking to the first;
  (ii) one still, the other moving and rebounding from the first;
  (iii) both still and 'exploded' apart by means of a spring originally compressed between them;
  (iv) both moving together and 'exploded' apart;
  (v) both moving, with collisions in which they stick together or rebound.

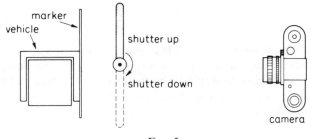

FIG. 5e

When vehicles are found to travel twice along the same part of track, the photographic record may become confused. It is then best to use a long shutter which covers up one part of the markers as they travel together, and is then flipped over at the collision to reveal another part as they move apart. See Fig. 5e.

25

# EXPERIMENT 6

# Materials: Breaking Strength and Fracture

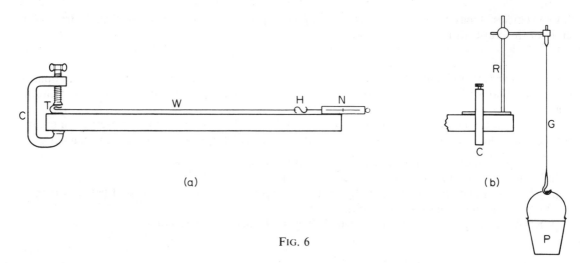

(a)　　　　　　　　　　　　　　　　(b)

FIG. 6

## APPARATUS

Copper wire, 32 s.w.g.; stainless steel wire, 44 s.w.g. (see *Caution* later); soft glass rod about 0·1 m long, 5 mm diameter; rubber bands (assorted); S-hook; retort stand and clamp; G-clamp; adhesive PVC tape; spring balances, 0–10 N, 0–50 N; plastic bucket *or* weight hanger and about ten 1 kg masses; standard wire gauge tables *or* micrometer screw-gauge; metre rule; hand lens *or* low-power microscope; bunsen burner and asbestos mat.

## METHOD

### General

In the experiments which follow, each of the materials copper, steel, glass and rubber will be stretched. The method for stretching the metals is different from that for glass and rubber. You will observe roughly how far each stretches before it breaks, the tension at fracture and the nature of the fracture.

### Experiment: Copper and Steel

1. Stretch the thin wires W as shown in Fig. 6(a). Attach one end T of the wire firmly to the bench; do this by folding a loop of the wire inside a folded-over piece of sticky PVC insulating tape and clamping this 'sandwich' firmly to the bench with a G-clamp C. Attach the other end of the wire, making it about 1 m long, to an S-hook H, by wrapping the wire several times round the hook and then round the wire itself.

2. Record the initial length of the wire. Stretch the wires *slowly*, using a spring balance N to pull on the S-hook. Find out if, for small extensions, the wire returns to its original length. Then, as the stretching is increased, note whether the wire is *permanently* stretched at larger extensions. Measure the extension of the wire approximately when it finally breaks (*Care*—see Caution below), using a metre rule laid on the bench.

3. Using a hand lens or low-power microscope, look at the two fractured ends of the wire and make a sketch of what you see. Look for any narrowing of the wire and for a rough region of fracture.

(*Caution.* Do *not* use appreciably thicker wire than specified, otherwise there is a danger that the

26

wire will whip back sharply when it fractures, the fast-moving end then being liable to cut someone, or worse, damage an eye.)

### Rubber and Glass

1. Use the vertical arrangement shown in Fig. 6(*b*). Here the sample G is attached to a retort stand and clamp R, itself fixed to the bench with a G-clamp C. Slotted masses can be used to load the sample at the lower end, or a plastic bucket P, slowly filled with water, will serve instead (Fig. 6(*b*)). Loads up to 10 kg may be needed; so in either case, make sure that the masses or bucket is only a few centimetres from the floor when fracture occurs. You will be expected (*a*) to record the load in kg at fracture and (*b*) to observe the fracture.

2. The *rubber bands* can simply be looped round the retort clamp and a hook attached to the load. Measure and record the width and thickness of the band before you load up.

The *glass* must be drawn into a thin fibre. To do this, first heat the glass rod at one end and bend it to form a hook (Fig. 6(*b*)). Allow the rod to cool. Now clamp the glass rod at its top end so that it is vertical, and attach a load of about 1 kg to the lower end with a string. Support the load and the bottom of the rod in one hand, and heat the middle of the rod until it is red-hot. Then let the load fall to the floor on to an asbestos mat—the glass will then be drawn into a fine fibre G, as shown in Fig. 6(*b*).

Now load up and fracture the fibre at once, while it is still clean—avoid touching the fibre, which would introduce cracks invisible to the eye and weaken the glass.

3. Look at the fractured ends of the rubber and glass with a hand lens or microscope. Also, measure and record the approximate extension at fracture. In the case of the glass you will probably not be able to observe any extension but it should be possible to say that it is less than the smallest amount you could have observed, so obtaining an 'upper limit'.

MEASUREMENTS

| Material | Original length /m | Extension at fracture /m | Original diameter /m | Tension at fracture /N |
|---|---|---|---|---|
| Copper<br>Steel<br>Rubber, first sample<br><br>. .<br><br>Glass, first sample<br><br>. . | | | | |

(*Experiment continued overleaf*)

## 6. (*Continued*)

CALCULATIONS

*Breaking stress.* Calculate the breaking stress, which is

$$\frac{\text{tension at fracture (N)}}{\text{cross-sectional area (m}^2)}$$

It is suggested that you use the *original* cross-section, not the cross-section at fracture. The stress is then called the 'engineering stress'. Note whether values obtained for different samples are roughly constant or not (they need not be constant).

Engineering stress at fracture:

$$\text{Copper} = \text{.. N m}^{-2}. \qquad \text{Steel} = \text{.. N m}^{-2}$$
$$\text{Rubber} = \text{.. N m}^{-2}. \qquad \text{Glass} = \text{.. N m}^{-2}$$

*Breaking strain.* Work out, even if roughly, the breaking strain, which is the percentage extension at fracture:

$$\text{Breaking strain} = \frac{\text{extension at fracture}}{\text{original length}} \times 100\%$$

You may give the value as a range, say from $0 \cdot 1\%$ to $0 \cdot 5\%$, or $50\%$ to $100\%$.

$$\text{Copper—from .. \% to .. \%.} \qquad \text{Steel —from .. \% to .. \%}$$
$$\text{Rubber—from .. \% to .. \%.} \qquad \text{Glass—from .. \% to .. \%}$$

## OBSERVATIONS AND CONCLUSIONS

Write an account of the differences in behaviour of the materials as they are stretched to fracture. State if there was a permanent extension, indicating *plastic yielding*, at any stage.

Decide if the fractured ends broke cleanly, so that one would fit into the other if they were put into contact, which indicates *brittle fracture*. Look also for evidence of *necking* before fracture, that is, a narrowing down of the material at one place, where ultimately it breaks.

In your account, contrast the various values of the stress at fracture, listing the materials in order of strength if possible. Note any materials for which a reliable constant value of this quantity cannot be obtained. Contrast also the differing breaking strains, or percentage extensions at fracture, and say what you can about the reasons for the differences.

# EXPERIMENT 7

# Measurement of the Young Modulus for a Wire

## APPARATUS

Two long steel wires X, Y, centimetre (M) and vernier (V) scales, kilogram masses to 10 kg and attachment, heavy mass A, metre rule, micrometer gauge.

## METHOD

Suspend the wires X, Y from beams or hooks in the ceiling, and attach the scales S, V. Place the heavy mass A on one wire to remove any kinks. With a load M such as 2 kg on the other wire, observe the reading of the scales. Add loads in steps of 2 kg to 12 kg, noting the scale readings on each occasion. (*Note.* Before taking readings, it may be necessary to load and unload the wire a few times.)

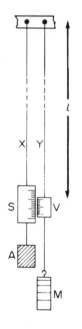

FIG. 7a

Remove the load in steps of 2 kg, observing the scale readings each time. (If the readings are much different from those when loading, the experiment should be repeated again with different wires, or with the kinks in the wires removed.) Measure the diameter of the wire, in two perpendicular directions, at three different places on the wire with a micrometer screw-gauge. Measure the length of the wire, *l*, from the ceiling to the beginning of the vernier scale.

## GRAPH

From the table of measurements, calculate the extensions for loads M of 2, 4, 6, 8 and 10 kg respectively, shown by $(b-a)$, $(c-a)$, $(d-a)$, $(e-a)$ and $(f-a)$ in the table.

Plot a graph of tensile force $F$ in newton against extension $e$ in metre. (Use $F=Mg$, where $g=$

30

$9.8\,\mathrm{N\,kg^{-1}}$). Draw the best straight line from the origin which passes through all the points (Fig. 7b). Measure the gradient $a/b$ of the line.

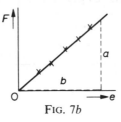

FIG. 7b

MEASUREMENTS

| Load /kg | Reading on scale /mm | Extension /mm |
|---|---|---|
| 0 | $a$ | |
| 2 | $b$ | $b-a$ |
| 4 | $c$ | $c-a$ |
| 6 | $d$ | $d-a$ |
| 8 | $e$ | $e-a$ |
| 10 | $f$ | $f-a$ |

Diameter of wire $d$ (6 values) = .. mm = .. m. Average $d$ = .. m
Length of wire, $l$              = .. m.

CALCULATION

$$\text{Young modulus, } E = \frac{F/A}{e/l} = \frac{F}{e} \times \frac{l}{A}$$

$$= \text{gradient} \times \frac{l}{\pi d^2/4}$$

From the gradient, $a/b$ in Fig. 7b, and $l$ and $d$ in metre, calculate $E$.

CONCLUSION

The value of the Young modulus for the wire = .. $\mathrm{N\,m^{-2}}$.

ERRORS

Errors occur in (1) reading the vernier, (2) the load, (3) the length of the wire. The percentage errors in (2) and (3) should be small, but the error (1) in the extension may be relatively large. The percentage error in $d^2$ will also be important.

ORDER OF ACCURACY

The order of accuracy can be found from the maximum or minimum gradient of the lines passing through the points, as explained on p. 9.

NOTE

From the measurements, the average extension $e$ for a load of 6 kg can be found from the average of the three values $(d-a)$, $(e-b)$ and $(f-c)$. See Table. The Young modulus can then be calculated from

$$E = \frac{F \cdot l}{A \cdot e} = \frac{6g}{\pi d^2/4} \times \frac{l}{e} = .. \ \mathrm{N\,m^{-2}}$$

31

# EXPERIMENT 8  Measurement of the Coefficients of Static and Dynamic Friction

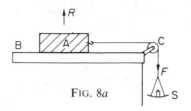

FIG. 8a

## COEFFICIENT OF STATIC FRICTION

APPARATUS

Wooden block A with a hook attached, a plane piece of wood B with a grooved wheel C at one end, scale-pan S, light string, weights, boxes of weights, spring balance.

METHOD

Weigh the block A and the scale-pan S on the spring balance. Attach the scale-pan to the hook of A by light string passing round the wheel C. Mark the position of A on the board B with pencil. Then gently add increasing weights to S until A just begins to slide. Record the total weight in S. Now increase the reaction of B by placing a known weight on A, and by adding increasing weights to S, again record the total weight in S when A begins to slip. Repeat for two more increasing weights on A, returning the block A to its original place on B each time.

MEASUREMENTS

Weight of scale-pan = .. N
Weight of block A = .. N

| Normal reaction, $R$/N | Weight in scale-pan on slipping/N | Limiting frictional force, $F$/N |
|---|---|---|
| | | |

CALCULATION

The limiting frictional force, $F$ = weight in scale-pan when A slips + weight of scale-pan.
Normal reaction, $R$ = weight of A + other weights on A.

GRAPH

Plot $F$ v. $R$ (Fig. 8b).
The gradient, $a/b = \mu = $ ...

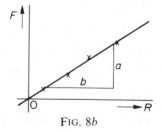

FIG. 8b

CONCLUSION

The coefficient of static friction between block and plane at the place concerned is ...

ERRORS AND ORDER OF ACCURACY

Draw the lines with the least and greatest slopes which just agree with the plotted points. Find the error in $\mu$ from the variation in slope.

# COEFFICIENT OF DYNAMIC FRICTION

APPARATUS

As before.

METHOD

With the apparatus shown in Fig. 8a, place a weight on S and give A a slight push towards C. Add increasing weights to S, giving A a slight push each time. At some stage, A will be found to continue moving with a steady, small velocity. Record the corresponding weight in the scale-pan S. Now increase the reaction of B by adding weights to A, and repeat. Repeat for two more weights on A, returning the block to its original place on B each time.

MEASUREMENTS

Weight of scale-pan = .. N
Weight of block A = .. N

| Normal reaction, $R$/N | Weight in scale-pan on moving A/N | Frictional force, $F'$/N |
|---|---|---|
|  |  |  |

CALCULATION

The frictional force, $F'$ = weight in scale-pan when A moves + weight of pan.
  Normal reaction, $R$ = weight of A + other weights on A.

GRAPH

Plot $F'$ v. $R$ (Fig. 8c).
  The gradient, $a/b = \mu' = \ldots$

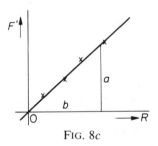

FIG. 8c

CONCLUSION

The coefficient of dynamic friction is ...

ERRORS AND ORDER OF ACCURACY

See previous page.

# EXPERIMENT 9

## Measurement of Viscosity by Stokes' Law

FIG. 9a

### APPARATUS

Measuring cylinder A, glycerine, stop-watch, small steel ball-bearings C of varying diameter, micrometer screw-gauge, metre rule, hydrometer, thermometer 0–100 °C.

### METHOD

Fill the measuring cylinder A with glycerine, and drop in the largest ball-bearing. Fix a mark X (sticky label is suitable) well below the top of the liquid, so that the bearing reaches a steady velocity by the time it reaches X. Fix a second mark Y near the bottom of the cylinder, and measure the distance $l$ cm between X and Y.

Time the fall of ball-bearings of varying diameter between the marks X and Y, having measured the diameter of each in two perpendicular directions with the micrometer screw-gauge. Note the zero error of the micrometer.

Measure the density of the glycerine with the hydrometer, and its temperature.

### MEASUREMENTS

Zero error of micrometer gauge = .. mm
Distance XY = $l$  = .. mm = .. × $10^{-3}$ m

| Micrometer readings /mm | | Average diameter /mm | Time of fall for $XY$ $t$/s | Terminal velocity $v$ /ms$^{-1}$ | $a^2$ (Radius$^2$) /m$^2$ |
|---|---|---|---|---|---|
|  |  |  |  |  |  |
|  |  |  |  |  |  |
|  |  |  |  |  |  |
|  |  |  |  |  |  |

Density of glycerine $\sigma$  = .. kg m$^{-3}$
Density of steel $\rho$  = .. kg m$^{-3}$ (from physical tables)
Temperature of glycerine = .. °C

34

## CALCULATION

Calculate the average diameter of each ball-bearing, the square of the radius, $a^2$, in metre$^2$ and the average terminal velocity $v$ for each bearing in metre second$^{-1}$; enter the results in the table of measurements.

## GRAPH

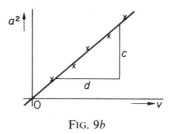

FIG. 9b

Plot a graph of $a^2$ against $v$, and draw the best line passing through the origin (Fig. 9b). Calculate the gradient, $c/d$.

## CALCULATION OF VISCOSITY

The terminal velocity $v$ is such that the apparent weight $\frac{4}{3}\pi a^3(\rho - \sigma)g$ is equal to the viscous drag $6\pi\eta av$. Thus the viscosity $\eta$ is given by:

$$\eta = \tfrac{2}{9}g(\rho - \sigma)\frac{a^2}{v} = \tfrac{2}{9}g(\rho - \sigma) \times (c/d) = \; ..$$

In SI units, $g = 9\cdot8\,\mathrm{m\,s^{-2}}$, $\rho$ and $\sigma$ are in $\mathrm{kg\,m^{-3}}$, $c$ is in m$^2$, and $d$ is in $\mathrm{m\,s^{-1}}$.

## CONCLUSION

The viscosity of glycerine at .. °C was .. $\mathrm{N\,s\,m^{-2}}$.

## ERRORS. ORDER OF ACCURACY

Find the error in the slope $c/d$ by drawing the line of greatest or least slope, passing near a majority of plotted points.

The max. percentage error in the viscosity is given by:

$$\frac{\delta\eta}{\eta} \times 100\% = \left[ \frac{\delta(c/d)}{(c/d)} \right] \times 100\%$$

neglecting errors in the densities.

# EXPERIMENT 10

# Measurement of Viscosity of Water by Capillary Tube Flow

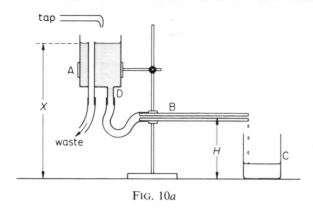

FIG. 10a

## APPARATUS

Constant-head apparatus A, capillary tube B of about 0·4 mm diameter, two clamps and stands, beaker C, stop-watch, rubber tubing for connecting A to B and for the water waste, half-metre rule, mercury, watch-glass, thermometer 0–100 °C, spirit level.

## METHOD

Place A below the water tap, connect rubber tubing to the waste outlet of A so that water runs away to the sink, and connect the other (constant-level) outlet D to the capillary tube B. Using the spirit level, clamp B horizontally at a distance H from the level of the bench. Turn on the water supply so that the water drips at a slow rate from the open end of B (a small spot of grease on the underside of the tube near the end will prevent water from running back beneath the tube). Start the stop-watch, and collect the flow in a dried, weighed beaker C, over a period of about 5 min. Weigh the water and beaker. Repeat the experiment for other values of H, varying the rate of flow as much as possible without allowing the flow to become rapid or turbulent. Measure the water temperature at intervals throughout the experiment. Remove the capillary tube and measure its length *l*.

Dry the capillary tube A, draw up a thread of mercury, and measure its length L. Weigh the watch-glass, expel the mercury into it, and reweigh.

## MEASUREMENTS

Mass of beaker　　　　　= .. g
Length of capillary tube $l$ = .. mm = .. m
Temperature of water　　= .. °C

| Height $H$ /m | Time /s | Mass of water + beaker /g | Volume of water /cm$^3$ | Volume per second /cm$^3$ s$^{-1}$ |
|---|---|---|---|---|
|  |  |  |  |  |

*Radius of Capillary Tube.* Length of mercury thread, $L$ = .. mm = .. m
Mass of mercury, $m$　　　　　= .. g = .. kg

36

GRAPH

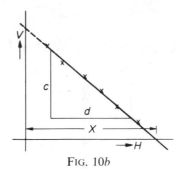

FIG. 10b

If the constant water level in A is $X$ m above the bench the head of water is $(X-H)$ m. Thus a graph of the volume of water per second $(V)$ against $H$ will be a straight line, by Poiseuille's formula:

$$V=\frac{\pi pa^4}{8\eta l}=\pi\frac{(X-H)\rho_w ga^4}{8\eta l} \qquad \ldots \ldots \ldots \ldots \ldots \quad (1)$$

where the density of water $\rho_w$ is $1000\,\text{kg m}^{-3}$.
Measure the slope $c/d$ (see Fig. 10b) of the 'best' line, changing $c$ to metre$^3$ second$^{-1}$ and $d$ to metre. From (1):

$$\text{Slope}=\frac{c}{d}=\frac{\pi g a^4}{8\eta l}. \qquad \ldots \ldots \ldots \ldots \ldots \quad (2)$$

CALCULATION

Calculate the radius $a$ of the capillary tube, using $m$ in kilogram, $L$ in metre, and density of mercury, $\rho=13\,600\,\text{kg m}^{-3}$:

$$a=\sqrt{\frac{m}{\pi.L\times 13\,600}}= \ldots\ \text{m}$$

Thus, from equation (2), the viscosity $\eta$ is given by:

$$\eta=\frac{\pi\rho_w g a^4}{8lc/d}= \ldots\ \text{N s m}^{-2}$$

CONCLUSION

The viscosity of water at .. °C is .. $\text{N s m}^{-2}$.

ERRORS

1. Estimate the error $\delta(c/d)$ in the gradient of the line by drawing the line whose slope differs as much as possible from the 'best' line, while still passing close to a majority of the points.

2. The percentage error in $a^4$ is $4\dfrac{\delta a}{a}\times 100\%$ where the error $\dfrac{\delta a}{a}=\dfrac{1}{2}\left(\dfrac{\delta m}{m}+\dfrac{\delta L}{L}\right)$. Compare this error with that incurred using a travelling microscope to measure the radius directly.

3. Error is made in measuring the length $l$ of the capillary tube.

ORDER OF ACCURACY

From the formula for the viscosity, expressing the radius $a$ in terms of the measured quantities, the maximum percentage error in the viscosity is:

$$\frac{\delta\eta}{\eta}\times 100\%=\left[\frac{\delta(c/d)}{c/d}+\frac{\delta l}{l}+2\left(\frac{\delta m}{m}+\frac{\delta L}{L}\right)\right]\times 100\%$$

37

# EXPERIMENT 11

# Measurement of Surface Tension of Water by Capillary Tube

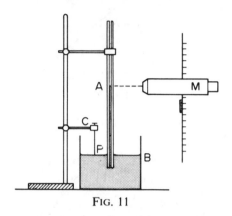

FIG. 11

## APPARATUS

Capillary tube A, travelling microscope M, beaker B, cork C with pin P, clamps and stand, thermometer, file for cutting capillary tube.

## METHOD

Do the experiment in a well-lit place, for example, near a window. Clean the capillary tube A with some dilute caustic soda and wash out repeatedly with distilled water. Fill the beaker B with water and measure its temperature. Now push the capillary tube A into the water so that the inside is wet, then raise the capillary tube so that the water inside reaches a level higher than the outside level of water in B, check that the level falls back to a constant position after being drawn up the tube, and fix the tube in a clamp. Push the pin P through the cork C, place the cork in a clamp as shown, and arrange the tip of the pin to just touch the water surface in B well away from the tube. This can be done accurately by means of the image of P in the water. Now focus the travelling microscope M on the meniscus of the water in A, and move the microscope until the horizontal crosswire is tangential to the bottom of the meniscus, which is seen inverted in M. If there is any difficulty in focusing M, hold a piece of paper on the glass at A and focus M on the paper first, as a guide. Read the travelling microscope. Mark the position of the meniscus A on the capillary tube with a pen or sellotape. Now carefully remove the capillary tube from B, then remove B, and focus the microscope on the tip of P; read the microscope vernier.

With the aid of a file, cut through the capillary tube at the position of the meniscus A. Measure two diameters at right angles at A by means of the travelling miscroscope.

## MEASUREMENTS

Temperature of water   = .. °C
Meniscus reading on microscope = .. mm
Pin reading on microscope   = .. mm
Diameter of capillary tube at A   = .., .. mm

38

## CALCULATION

$$\text{Height of water column } h \qquad = \; .. \; \text{mm} = \; .. \; \text{m}$$
$$\text{Average radius of capillary tube } r = \; .. \; \text{mm} = \; .. \; \text{m}$$

For water,
$$\gamma = \frac{rh\rho g}{2} = \; .. \; \text{N m}^{-1}$$

where $h$ and $l$ are in m, $\rho$ is $1000 \, \text{kg m}^{-3}$ and $g = 9 \cdot 8 \, \text{m s}^{-2}$.

## CONCLUSION

The surface tension of water at .. °C was .. $\text{N m}^{-1}$.

## ERRORS

The error in the capillary rise $h$ is due to (i) errors in the two vernier readings, (ii) error in setting the crosswires on the meniscus or pin, (iii) error in setting the pin at the water surface. Similar errors to (i) and (ii) occur in measuring the radius.

## ORDER OF ACCURACY

From the formula for $\gamma$ in terms of measured quantities,

$$\text{max. \% error in } \gamma = \left( \frac{\delta h}{h} + \frac{\delta r}{r} \right) \times 100\%$$

## NOTE

The mean radius $r$ of the capillary tube can be found by measuring the length $l$ and mass $m$ of a mercury thread drawn into it. The radius $r$ in the surface tension formula, however, is the radius of the tube at the meniscus and not the mean radius.

# PART THREE

# Heat

# EXPERIMENT 12

## Measurement of Temperature by a Thermistor Thermometer

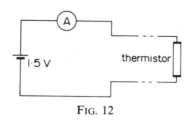

FIG. 12

APPARATUS

Dry cell 1·5 V, 0–10 mA meter, thermistor (TH-B12 from R.S. Components Ltd.) glued with Araldite to end of pyrex tube. Ethanoic acid, ice, stop-watch, mercury thermometer − 10 °C to 110 °C, beaker, test-tube.

METHOD

1. Connect the cell, meter and thermistor in series (Fig. 12). Place the thermistor and mercury thermometer in melting ice in a beaker. Record the meter reading at 0 °C.

2. Remove the ice, add water if necessary to cover the thermistor and thermometer bulb. Warm the water and take readings of the meter at intervals of about 10 °C, as measured with the mercury thermometer, until the water boils. Record the meter reading at the temperature of the boiling water, which we assume is 100 °C.

3. Remove the thermistor. When it is cool, place it under your arm near the body and record the meter reading when it is steady.

4. Put a few cm³ of ethanoic acid in a test-tube, place the thermistor in the acid so that it is completely covered, and put the test-tube in a beaker of melting ice. (You may need to melt the ethanoic acid before putting it into the test-tube as the melting-point is near room temperature.) When the acid has frozen, remove the test-tube from the melting ice and allow it to warm up again. At the same time, take readings of the meter every 30 s until the acid has all melted. (You may need to put the frozen acid into a beaker of warm water to help it start to melt.)

MEASUREMENTS

(a)

| Temperature /°C | Meter current /mA |
| --- | --- |
|  |  |

(b) When placed under the arm, meter reading = .. mA
(c) For ethanoic acid:

| Time /min | Meter current /mA |
| --- | --- |
|  |  |

## CALCULATIONS

1. Using the meter readings at the upper fixed point (100 °C) and lower fixed point (0 °C), plot these two points on a graph of meter reading against temperature in °C and draw the straight line between them.

From your *measurements* above, note the meter reading $I$ corresponding to 50 °C. From your graph, note the temperature corresponding to this meter reading. Explain the discrepancy between the two temperatures.

2. A temperature $\theta_{th}$ on the thermistor centigrade scale may be defined by the relation

$$\theta_{th} = \frac{I_\theta - I_0}{I_{100} - I_0} \times 100\,°C$$

where $I_{100}$ and $I_0$ are the currents at 100 °C and 0 °C respectively and $I_\theta$ is the current at the temperature $\theta_{th}$. Use this relation to calculate, from the readings in (*a*), the temperature on the thermistor temperature scale corresponding to 50 °C on the mercury-in-glass temperature scale.

3. Plot a graph of meter current against temperature on the thermistor temperature scale from your readings of the current $I$ in the table in (*a*)—start with 0 °C and finish with 100 °C.

4. Using the meter reading when the thermistor was under your arm, calculate this temperature on the thermistor scale from the above relation for $\theta_{th}$.

5. Plot a graph of the meter reading against time when the thermistor was inside the ethanoic acid. From the graph, determine the melting-point of the acid on the thermistor temperature scale.

## CONCLUSIONS

The temperature on the thermistor temperature scale corresponding to 50 °C on the mercury-in-glass temperature scale is . . °C.

The melting-point of ... is . . °C.

## QUESTIONS

(*a*) From your meter and temperature readings, does the thermistor resistance (i) increase or decrease with temperature rise, (ii) change more between 0 °C and 10 °C than between 90 °C and 100 °C?

(*b*) On what temperature scale did you calculate the temperature under your arm? How does this compare with normal body temperature?

(*c*) On what temperature scale did you determine the melting-point of ethanoic acid? Look up the melting-point in standard Tables and comment on how well your value compares with the accepted value.

# EXPERIMENT 13

## Measurement of Temperature and Melting-point by Thermocouple

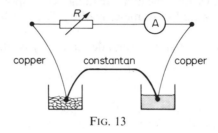

FIG. 13

Copper–constantan thermocouple, microammeter (Scalamp, Pye or Edspot galvanometer), resistance box, 0–1000 $\Omega$, naphthalene, beakers, ice.

METHOD

1. *Lower and upper fixed points determination.* Connect the thermocouple, resistance box R and meter A in series (Fig. 13). Place one junction of the thermocouple in melting ice and the other in tap water. Check that the pointer of the meter moves the correct way across the scale. If not, reverse the connections to the meter.
2. Now place both junctions of the thermocouple in melting ice. Using the 'zero set' of the meter, adjust the pointer so that it is at the extreme left of the scale. This is the meter reading at the lower fixed point of the thermocouple thermometer.
3. Place the 'hot' junction of the thermocouple back in the beaker containing the water and heat the water until it is boiling. Adjust the resistance in R so that the deflection at this upper fixed point is some large convenient value, easily divisible by 100, from the lower fixed point.
4. *Temperature measurement of melting-point.* Put a few cm³ of naphthalene in a test-tube and melt it by holding the tube in a clamp in water brought to boiling. When the naphthalene has all melted, place the 'hot' junction of the thermocouple inside it and allow the test-tube to cool in air, with the 'cold' junction in melting ice—clamp the thermocouple so that it does not move. Take readings of the meter every 30 s until the naphthalene has all solidified and the readings have remained constant for a period.

MEASUREMENTS

*Fixed points*

Meter reading at lower fixed point = … divisions
Meter reading at upper fixed point = … divisions

*Melting-point of naphthalene*

| Time /min | Meter reading /divisions |
|-----------|--------------------------|
|           |                          |

## CALCULATION

On the copper–constantan thermocouple centigrade scale, the temperature $\theta_c$ is defined by the relation

$$\theta_c = \frac{E_\theta - E_0}{E_{100} - E_0} \times 100\,°C$$

where $E_\theta$, $E_{100}$ and $E_0$ are the respective e.m.f.s of the thermocouple at the temperature $\theta_c$ and the upper and lower fixed points. The current $I$ flowing in the meter is proportional to the thermocouple e.m.f. $E$, if the total resistance in the circuit remains constant during the experiment. So if $I_\theta$, $I_{100}$ and $I_0$ are the respective currents at temperature and the upper and lower fixed points, then, from above,

$$\theta_c = \frac{I_\theta - I_0}{I_{100} - I_0} \times 100\,°C$$

## GRAPH

Plot a graph of the meter reading $I$ against time $t$ for the naphthalene cooling. From the graph, determine the meter reading $I$ at the melting-point of naphthalene. Use the above relation to obtain the melting-point $\theta_c$ on the thermocouple scale.

## QUESTIONS

(a) Does the value of the resistance in the resistance box R affect the meter reading at the lower fixed point?

(b) What feature of the cooling curve is linked with the melting-point and why?

(c) What is the meter reading when the temperature on the thermocouple temperature scale is 50 °C?

(d) Look up the melting-point of naphthalene in standard Tables and comment on the agreement with the value you obtained.

# EXPERIMENT 14

## Specific Heat Capacity of Metal by Electrical Method

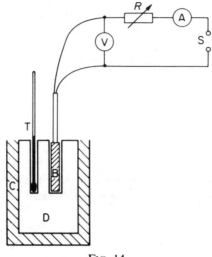

FIG. 14

APPARATUS

Metal (aluminium) block D with holes for immersion heater (heating coil) B and thermometer T (Griffin & George *or* Philip Harris); lagging C; suitable supply S, voltmeter V, ammeter A, rheostat R; stop-watch.

METHOD

1. Insert the immersion heater B and thermometer T in the metal block (use a little vaseline to improve thermal contact between the thermometer bulb and the metal) (Fig. 14).
2. Connect the electric circuit shown to the heater B. Use suitable high values of $I$ and $V$.
3. Record the initial steady temperature on T. Then switch on the current, simultaneously start the stop-watch, and record the temperature on T at equal intervals such as $\frac{1}{2}$ min or 1 min. During the heating, keep the current and p.d. constant by means of the rheostat $R$.
4. After a suitable temperature rise has been obtained, switch off the current, and note the time $t$ of heating. As the temperature of the metal continues to rise, record (*a*) its final maximum temperature, and then (*b*) the lower temperature reached 1 min later.
5. Record the mass of the metal block. Repeat the experiment with values of $V$ and $I$ a little lower than before.

46

## MEASUREMENTS

| | Expt. 1 | 2 |
|---|---|---|
| Ammeter reading, $I$ /A | | |
| Voltmeter reading, $V$ /V | | |
| Initial temp. of metal, $\theta_1$ /°C | | |
| Maximum temp. of metal, $\theta_2$ /°C | | |
| Time of heating, $t$ /s | | |
| Temp. reached from $\theta_2$ in 1 min, $\theta_3$ /°C | | |
| Mass of metal block, $m$ /kg | | |

## CALCULATION

If $c$ is the specific heat capacity of the metal, then, assuming no heat losses,

$$\text{energy supplied, } IVt = mc(\theta_2 - \theta_1)$$

Calculate $c$ from this expression in the two experiments.

## CONCLUSION

The specific heat capacity of ... is ... $J\,kg^{-1}\,K^{-1}$.

## ERRORS. ORDER OF ACCURACY

1. Some heat is lost to the surroundings.
2. The ammeter and voltmeter have errors.
3. The thermometer temperature is assumed to be that of the metal block.

The percentage error in $c$ is the sum of the percentage errors in $I$, $V$, $t$, $m$ and the temperature rise. The errors in $t$ and $m$ are relatively small and may be neglected; those in $I$ and $V$ may be of the order of 2% each.

If there is error due to some *energy transfer to the surroundings as heat*,

(1) Will this make your result for $c$ too large or too small?

(2) Suggest one or more ways of estimating how much heat is lost.

(3) Using the temperature fall in 1 minute from the maximum measured in your experiment, $\theta_2 - \theta_3$, make some estimate of the error in your answer for $c$ due to heat loss to the surroundings.

# EXPERIMENT 15

# Measurement of Specific Heat Capacity of Oil by Electrical Method

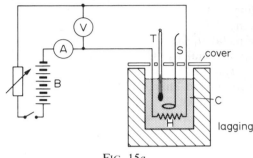

FIG. 15a

## APPARATUS

Well-lagged copper calorimeter C, sensitive thermometer T, copper stirrer S, heating coil H, voltmeter V, ammeter A, rheostat $R$, 12 V accumulator B, switch, stop-watch, Castrolite oil.

## METHOD

Connect the circuit as shown in the diagram, including a rheostat $R$ of a suitable resistance. Weigh the empty calorimeter and stirrer, and then with sufficient oil added to cover the heating coil completely. Switch on the circuit, and check that the value of the current is suitable (about 2 A). Switch off, stir well, and take the initial temperature. Switch on again, and start the stop-watch. Take readings of the temperature every half-minute, stirring continually, and keeping the current constant using $R$, until the temperature has risen about 10 K *but not higher* (CARE). Switch off, but continue to read the temperature every half-minute until it has fallen several degrees. Record the maximum temperature reached. Record the readings of the ammeter $A$ and voltmeter $V$.

## MEASUREMENTS

Mass of calorimeter + stirrer $M$ = .. kg, Specific heat capacity $c$ = .. $\mathrm{J\,kg^{-1}\,K^{-1}}$
Mass of calorimeter + stirrer + oil = .. kg
$\therefore$ Mass of oil $m$ = .. kg

| Temperature /°C | | | | | | | | | | | | | | | | | |
|---|---|---|---|---|---|---|---|---|---|---|---|---|---|---|---|---|---|
| Time /s | | | | | | | | | | | | | | | | | |

Current $I$ = .. A      Initial water temperature = .. °C
P.d. across coil $V$ = .. V      Maximum temperature = .. °C

## CALCULATION

Heat supplied $= IVt$. The heat is in joule when $I$, $V$, $t$ are in ampere, volt, and second respectively.

Thus
$$(Mc + mc_1)\theta = IVt$$

where $m$ is the mass of oil of specific heat capacity $c_1$, $M$ is the mass of the calorimeter and stirrer, $\theta$ is the temperature rise (corrected for cooling), and $t$ is the time for which the current was switched on. With $m$ and $M$ in kg, calculate $c$.

48

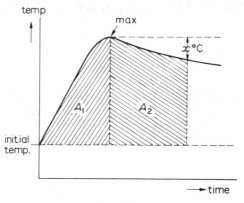

FIG. 15$b$

Plot a graph of temperature (°C) v. time (min). The 'cooling correction', $r$, is the temperature to be added to the final temperature observed, and can be calculated from the relation

$$\frac{r}{A_1}=\frac{x}{A_2} \quad \text{or} \quad r=\frac{A_1}{A_2}x$$

where $x$ is any temperature drop from the observed final temperature, and $A_1$, $A_2$ are the areas shown in Fig. 15$b$. The areas $A_1$, $A_2$ are obtained by counting squares. From the graph,

$$r=\frac{A_1}{A_2}x=\ \ ..\ \text{K}$$

## CONCLUSION

The specific heat capacity of the oil was found to be .. $J\,kg^{-1}\,K^{-1}$.

## ERRORS, ORDER OF ACCURACY

1. The masses may be determined to within 0·1% by weighing to the nearest tenth of a gram. No greater accuracy is required, as the other errors exceed this considerably.

2. The errors in the reading of the ammeter and voltmeter will depend upon the instruments used, and the size of their scales.

3. The error in the time is small, since the heating is fairly slow and $t$ is large.

4. The error in $\theta$ is the sum of the errors in measuring the initial and final temperatures and the cooling correction.

Neglecting the small percentage errors in $t$, $M$ and $m$, the maximum percentage error in $c_1$ is given approximately by:

$$\frac{\delta c_1}{c_1}\times 100\%=\left(\frac{\delta\theta}{\theta}+\frac{\delta I}{I}+\frac{\delta V}{V}\right)\times 100\%$$

## NOTE

($a$) The coil should have a resistance of about 5–6 $\Omega$, and should be well varnished.

($b$) The same method can be used to measure the specific heat capacity of water.

49

# EXPERIMENT 16

## Measurement of Specific Heat Capacity of Metal by Mechanical Energy Method

### APPARATUS

Mechanical energy to heat laboratory apparatus, with solid cylindrical metal calorimeter C mounted on a nylon insulating bush, crank A, short-range thermometer D, cord E attached to rubber band R, mass M of several kg, calipers, stop-watch.

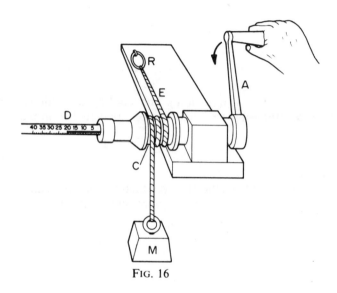

FIG. 16

### METHOD

1. Arrange the apparatus as shown in Fig. 16. With one end of the cord E attached to the band R, wind several turns round the calorimeter C and then attach a mass M to the other end. When the crank A is turned steadily, make sure the mass M hangs *freely* from the cord, so that the whole of its weight is supported.

2. Place the thermometer axially inside the hole in the metal calorimeter—smear the bulb with a little glycerine to improve thermal contact with the metal. Note the steady temperature of the metal.

3. Now start turning the crank A steadily at a rate such that M is supported completely by the cord and the rubber band is fairly slack. Simultaneously start the stop-watch and begin counting the number of revolutions as you turn.

After a suitable temperature rise, of the order of 10 K, stop turning, note the number of revolutions, and observe the final maximum temperature on the thermometer and the time taken to reach it—the temperature continues to rise after you have stopped turning. Then as the calorimeter cools observe the time for the temperature to fall by 1 K from the maximum.

Record or measure the mass of the metal calorimeter and measure the diameter of the calorimeter in several places with the calipers.

4. Repeat the experiment, this time obtaining a greater temperature rise.

## MEASUREMENTS

Initial calorimeter temp., $\theta_1$ .. °C
Final calorimeter temp., $\theta_2$ .. °C
No. of revolutions, $n$ ..
Mass of weight, $M$ .. kg
Mass of calorimeter, $m$ .. kg
Diameter of calorimeter, $d$ .. m
Time of rotation, $t$ .. s
Time for temp. to fall by 1 K .. s

## CALCULATION

*Cooling correction.* Calculate the rate of fall of temperature per second at the maximum temperature. Assuming a uniform temperature rise of the calorimeter during heating,

$$\text{cooling correction, } \theta_c = \frac{\text{rate of cooling}}{2} \times t$$

*Calculation for c.* Work done against frictional force between cord and metal $= Mg \times \pi dn$

$$\text{Heat produced} = mc(\theta_2 + \theta_c - \theta_1)$$

where $c$ is the specific heat capacity of the metal of the calorimeter. Assuming all the mechanical energy is transferred as heat to the calorimeter,

$$mc(\theta_2 + \theta_c - \theta_1) = Mg \times \pi dn$$

Calculate $c$ from this relation.

## CONCLUSION

The specific heat capacity of the metal is .. $J\,kg^{-1}\,K^{-1}$.

## ERRORS

The frictional force between the cord and metal may vary during the rotation, in which case the frictional force is not constant. Further, (a) not all the mechanical energy may be transferred as heat to the calorimeter, (b) the correction for heat losses is approximate.

# EXPERIMENT 17

## Measurement of Specific Heat Capacity of Water by Continuous-flow Method

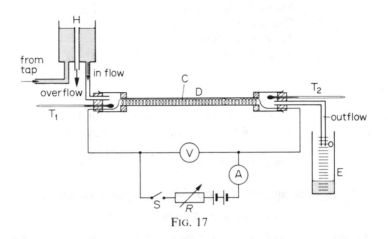

FIG. 17

APPARATUS

Ammeter A, voltmeter V, 12 V accumulator or d.c. supply, rheostat $R$, sensitive thermometers $T_1$, $T_2$, 'constant-head' apparatus H, 'constant-flow' apparatus D, switch S, stop-watch, measuring cylinder E.

METHOD

Connect the heating coil C in series with the battery, rheostat $R$, ammeter A, and switch S. Join the voltmeter V in parallel with the coil C. Make sure that the + and − terminals of the meters are correctly connected in the circuit. Switch on, and adjust the current $I$ to a suitable value, calculating the approximate number of joules per second being dissipated. Adjust the rate of flow by moving H or the outlet O, until a slow, steady flow is obtained. Measure this rate roughly, and calculate the expected temperature difference, which should not be less than, say, 4 °C. Clamp the outlet O in position, as moving it will affect the rate of flow. Allow the temperature difference to settle down to its steady value, reading $T_2$ when this state has been reached. Meanwhile, make several measurements of the rate of flow by collecting the outflow in a measuring cylinder (or weighed beaker) for a known time, not less than 30 s. The current $I$ must be kept constant throughout using $R$, and the readings of the ammeter and voltmeter should be checked at intervals.

Now reduce the current to about three-quarters of its former value, and decrease the rate of flow until the steady temperature difference is the *same* as in the previous part so that the rate of heat loss is the same. Measure the new rate of flow, and take the new ammeter and voltmeter readings.

| | 1st part | 2nd part |
|---|---|---|
| Current $I$/A | $I_1$ | $I_2$ |
| Voltage $V$/V | $V_1$ | $V_2$ |
| Mass of water collected/kg | | |
| Time/s | | |
| $\therefore$ Rate of flow/kg s$^{-1}$ | $m_1$ | $m_2$ |
| Inflow temperature/°C | | |
| Outflow temperature/°C | | |
| $\therefore$ Temperature difference/K | $\theta$ | $\theta$ |

CALCULATION

Rate of heat supplied $= I_1 V_1 = m_1 c_w \theta + h$, where $c_w$ is the specific heat capacity of water and $h$ is the rate of loss of heat. Since $h$ is the same in both parts:

$$I_2 V_2 = m_2 c_w \theta + h$$
$$\therefore \ I_1 V_1 - I_2 V_2 = (m_1 - m_2) c_w \theta$$

With $m_1$ and $m_2$ in kg s$^{-1}$, calculate $c_w$.    $\therefore \ c_w = \dfrac{I_1 V_1 - I_2 V_2}{(m_1 - m_2)\theta}$

CONCLUSION

The value of the specific heat capacity of water was .. J kg$^{-1}$ K$^{-1}$.

ERRORS

The error due to loss of heat is eliminated at the expense of further measurements, which are themselves subject to error.

1. Errors occur in reading the current and p.d., the errors depending on the meters employed.

2. The percentage error in measuring the rates of flow may be made quite small by collecting the water for as long as possible. Check that the rate of flow is not varying.

3. The error in the temperature difference may be quite large, being the sum of the errors in the inflow and outflow temperatures. Note whether the two thermometers agree initially, and if they do not, correct the temperature difference for any discrepancy.

ORDER OF ACCURACY

Assuming maximum, or minimum, errors in the values of current, p.d., mass and temperature change, a rough estimate of the order of accuracy can be obtained by substituting in the equation for $c_w$ values of $I_1 V_1 - I_2 V_2$, $m_1 - m_2$, and $\theta$ which are as high, or as low, as possible. This gives the highest (or lowest) value which $c_w$ could have. See also p. 57.

NOTE

The apparatus, if not available, may be constructed from about 30 cm of 0·5 cm bore glass tubing, containing a spiral coil of constantan, well varnished, of about 5–6 $\Omega$ resistance (i.e. 24 W at 12 V). End tubes of about 2 cm bore carry corks bored for the inflow and outflow tubes, thermometers, and the leads to the coils. A constricted outlet at O is advantageous.

# EXPERIMENT 18

## (i) Heat Capacity of a Metal Block, (ii) Specific Heat Capacity of Oil, by Mixtures

### APPARATUS

Large mass of metal (about 0·2 kg) A, beaker B, copper calorimeter C in insulating jacket D, copper stirrer E, tripod, gauze, burner, chemical balance, weights, oil (e.g. paraffin or Castrolite), thread, stopwatch, thermometer 0–100 °C.

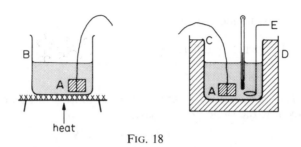

FIG. 18

### (i) HEAT CAPACITY OF METAL

### METHOD

Fill the beaker B with some water, place the metal A inside it, and boil the water. Meanwhile, weigh the calorimeter and stirrer, fill it about one-half with water, and reweigh. Take the temperature of the water. Take the temperature of the boiling water, and then quickly transfer A to the water in the calorimeter C. Observe the water temperature every 10 s until it reaches a maximum and then drops several degrees below the maximum reached.

### MEASUREMENTS

$$
\begin{aligned}
&\text{Mass of calorimeter} + \text{stirrer } m_1 \; (c_1 = \;.. \; \mathrm{J\,kg^{-1}\,K^{-1}}) &&= \;.. \; \mathrm{kg} \\
&\text{Mass of calorimeter} + \text{stirrer} + \text{water } m_1 + m &&= \;.. \; \mathrm{kg} \\
&\text{Initial water temperature } t_1 &&= \;.. \; \mathrm{^\circ C} \\
&\text{Final temperature observed} &&= \;.. \; \mathrm{^\circ C} \\
&\text{Final temperature, corrected for cooling } t_2 &&= \;.. \; \mathrm{^\circ C} \\
&\text{Temperature of boiling water } t &&= \;.. \; \mathrm{^\circ C}
\end{aligned}
$$

### COOLING CORRECTION

This may be obtained by a graphical method, as explained on p. 49. An alternative method is as follows: Suppose it took a time $x$ for the water to reach its final temperature when the hot metal was dropped in; then, approximately, the cooling correction is the temperature drop from the maximum temperature in a time $x/2$. Since a metal is a good conductor, it gives up its heat quickly, and the cooling correction may therefore be negligible.

## CALCULATION

Heat lost by metal = Heat gained by water and calorimeter + stirrer. If $C$ is the heat capacity of the metal, and $m$ the mass of water of specific heat capacity $c_W$ ($= 4200 \, \mathrm{J \, kg^{-1} \, K^{-1}}$), then

$$C \times (t - t_2) = (mc_W + m_1 c_1)(t_2 - t_1)$$

With $m$ and $m_1$ in kg, calculate $C$.

## CONCLUSION

The heat capacity of the metal was .. $\mathrm{J \, K^{-1}}$.

## ERRORS

(1) Heat lost by the hot metal on transferring it to the calorimeter; (2) some hot water is carried over with the metal; (3) observations of the temperature (e.g. $16.4 \pm 0.2 \, °\mathrm{C}$) and mass (e.g. $194.6 \pm 0.1 \times 10^{-3} \, \mathrm{kg}$).

## ORDER OF ACCURACY

See p. 10.

# (ii) SPECIFIC HEAT CAPACITY OF OIL

## METHOD

Add some water to the beaker, place the metal A inside it, and heat the water until it boils. Meanwhile, weigh the calorimeter, fill it about one-half with the oil, and reweigh. Observe the oil temperature. Take the temperature of the boiling water, and then quickly transfer A to the oil. Observe the time taken for the oil to reach its maximum temperature, and then find the temperature drop, $c$, in half this time. This is the cooling correction (see previous experiment).

## MEASUREMENTS

| | |
|---|---|
| Mass of calorimeter $m_1$ ($c_1 = \,.. \, \mathrm{J \, kg^{-1} \, K^{-1}}$) | $= \,..$ kg |
| Mass of calorimeter + oil $m_1 + m$ | $= \,..$ kg |
| Initial oil temperature $t_1$ | $= \,..$ °C |
| Final temperature, corrected for cooling $t_2$ | $= \,..$ °C |
| Temperature of boiling water $t$ | $= \,..$ °C |
| Heat capacity of metal ($C$)—from previous experiment | $= \,.. \, \mathrm{J \, K^{-1}}$ |

## CALCULATION

Heat loss by metal = Heat gained by oil and calorimeter. If $c$ is the oil's specific heat capacity and $m$ is the mass of the oil, then, with $m$ and $m_1$ in kg, calculate $c$ from

$$H \times (t - t_1) = (mc + m_1 c_1)(t_2 - t_1)$$

## CONCLUSION

The specific heat capacity of the oil was .. $\mathrm{J \, kg^{-1} \, k^{-1}}$.

## ERRORS AND ORDER OF ACCURACY

See previous experiment and p. 10.

# EXPERIMENT 19

## Specific Latent Heat of Vaporisation by Electrical Method

### APPARATUS

Flask F with surrounding vapour jacket and immersion heater H (Fig. 19a), *or* thermos flask arrangement T and heater H (Fig. 19b), condenser C. Suitable voltage supply S, ammeter A, voltmeter V, rheostat R. Three conical flasks and filter paper. Ethanol or other suitable liquid. Stop-watch.

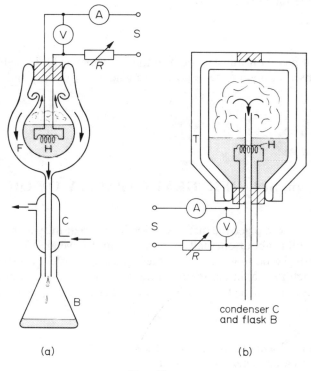

(a)                              (b)

FIG. 19

### METHOD

1. Fill the flask F with sufficient liquid to cover the heater H. Weigh two clean conical flasks, with filter paper covering the top each time. Place the unweighed conical flask B below the condenser C to collect condensed vapour (condensate).

2. Connect the supply S to the heater H, with the ammeter A, voltmeter V and rheostat R in the circuit as shown. Switch on the maximum power, then wait for the liquid to boil and for condensed vapour to collect in B. When this happens, adjust the rheostat R so that convenient high values of current and voltage are obtained, and record their values.

3. As soon as the condensate drips steadily, replace B by a preweighed conical flask and immediately start the stop-watch. Maintain steady values of current $I$ and voltage $V$ by means of R. After a suitable period $t$, making sure that the liquid covers the heater H throughout, weigh the flask and condensate with the filter paper on top as before. The mass of the condensed vapour can then be found.

4. Repeat the whole experiment with a steady lower current $I$ and voltage $V$, using the second preweighed conical flask.

## MEASUREMENTS

|                                   | Expt. 1 | Expt. 2 |
| --------------------------------- | ------- | ------- |
| Current, $I$/A                    | . .     | . .     |
| Voltage, $V$/V                    | . .     | . .     |
| Time, $t$/s                       | . .     | . .     |
| Mass of preweighed flask /kg      | . .     | . .     |
| Mass of flask + condensate /kg    | . .     | . .     |

## CALCULATION

Suppose $m_1$ and $m_2$ are the respective masses per second of condensed vapour, in $kg\,s^{-1}$, in experiments 1 and 2; $I_1 V_1$ and $I_2 V_2$ the respective powers in W supplied; and $h$ the heat lost per second in W by the whole apparatus to the surroundings when the vapour was produced (we assume $h$ to be the same in the two experiments since the jacket J is at the same temperature, the boiling-point of the liquid, each time). Then, if $l$ is the specific latent heat of vaporisation,

$$I_1 V_1 = m_1 l + h \qquad \ldots \ldots \ldots \ldots \quad (1)$$

$$I_2 V_2 = m_2 l + h \qquad \ldots \ldots \ldots \ldots \quad (2)$$

So
$$I_1 V_1 - I_2 V_2 = (m_1 - m_2)l \quad \ldots \ldots \ldots \ldots \quad (3)$$

Calculate $l$ from equation (1), *assuming h is relatively small compared with the power supplied and hence negligible*. For the use of equation (3), see *Order of Accuracy*.

## CONCLUSION

The specific latent heat of vaporisation of ... is .. $J\,kg^{-1}$.

## ERRORS

The percentage error in the power supplied, $IV$, is the sum of the percentage errors in $I$ and $V$. If the meters each have an error of 2%, the total error in $IV$ is 4%. The percentage errors in the mass and time are relatively small and may be neglected. Note that the vapour jacket outside the boiling liquid reduces considerably the heat losses from the liquid, although some heat $h$ is lost to the surroundings.

## ORDER OF ACCURACY

Although equation (3) eliminates $h$, which is unknown, the error in the measurement of the difference in power supplied, $I_1 V_1 - I_2 V_2$, is much more than the error in $I_1 V_1$, the power supplied in (1). So the determination of $l$ from equation (3) using school laboratory apparatus may not be so accurate as using equation (1).

Obtain the value of $l$ for the liquid from standard Tables. Compare it with your result using equation (1) and equation (3). Which equation gives the more accurate result? What percentage is $h$ of the power supplied in (1) and does this justify neglecting $h$ in this equation?

# EXPERIMENT 20

# Measurement of (i) Specific Latent Heat of Fusion of Ice, (ii) Specific Heat Capacity of Oil using Ice

## APPARATUS

Ice, blotting paper, calorimeter, and insulating jacket, stirrer, tripod, gauze, burner, thermometer 0–50 °C in 0·1 °C or 0·2 °C graduations, beaker, oil (e.g. paraffin or Castrolite), chemical balance, weights.

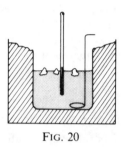

FIG. 20

## (i) SPECIFIC LATENT HEAT OF ICE

### METHOD

Warm water gently in a beaker until it is several degrees above the temperature of the surroundings. Weigh the calorimeter and stirrer, fill it one-third full with the warm water, and reweigh. Observe the initial water temperature when it is, say, 6 K above that of the surroundings. Now add small pieces of ice which have been carefully dried with blotting paper until the temperature of the water is the same number of degrees below room temperature as it was originally above room temperature. Be careful to stir until *all* the ice is melted before adding more ice. Now reweigh the calorimeter and its contents.

### MEASUREMENTS

Mass of calorimeter + stirrer $m_1$ ($c_1 = $ .. $\mathrm{J\,kg^{-1}\,K^{-1}}$) = .. kg
Mass of calorimeter + stirrer + water = .. kg
Mass of calorimeter + stirrer + water + ice = .. kg
Initial water temperature $t_1$ = .. °C
Final water temperature $t_2$ = .. °C

### CALCULATION

Mass of water $m_2$ ($c_W = 4200\,\mathrm{J\,kg^{-1}\,K^{-1}}$) = .. kg
Mass of ice $m$ = .. kg

Heat gained by ice in melting to water = heat lost by water and calorimeter

$$\therefore\ ml + mc_W(t_2 - 0) = (m_2 c_W + m_1 c_1)(t_1 - t_2)$$

### CONCLUSION

The latent heat of fusion of ice was .. $\mathrm{J\,kg^{-1}}$.

## ERRORS

(1) The ice is not perfectly dry, thus leading to error in the mass of ice, $m$. (2) Errors occur in the measurements of mass and temperature. (3) The net heat transfer between the surroundings and the water in the calorimeter is substantially eliminated by the above method.

## ORDER OF ACCURACY

Since $l = Q/m$, where $Q = (m_2 c_w + m_1 c_1)(t_2 - t_1) - mc_w t_2$, the percentage error in $l$ is the sum of the percentage errors in $Q$ and $m$. The percentage error in $m$ has a great effect on the value of $l$, since (a) $m$ is small and (b) a significant mass of water may be carried over with the ice if it is not perfectly dry.

# (ii) SPECIFIC HEAT CAPACITY OF OIL USING ICE

## METHOD

Warm some oil gently in a beaker until it is some degrees above the temperature of the surroundings. Weigh the calorimeter, fill it about one-third full of the warm oil, and reweigh. Take the temperature of the oil when it is about $6\,°C$ above the room temperature, and then add small *dry* pieces of ice until the temperature is the same number of degrees below the room temperature, stirring until all the ice is melted. Then reweigh the calorimeter and contents.

## MEASUREMENTS

$$\begin{aligned}
\text{Mass of calorimeter } m_1 \ (c_1 = \ .. \ \mathrm{J\,kg^{-1}\,K^{-1}}) &= \ .. \ \mathrm{kg} \\
\text{Mass of calorimeter} + \text{oil} &= \ .. \ \mathrm{kg} \\
\text{Mass of calorimeter} + \text{oil} + \text{ice} &= \ .. \ \mathrm{kg} \\
\text{Initial oil temperature } t_1 &= \ .. \ °C \\
\text{Final oil temperature } t_2 &= \ .. \ °C
\end{aligned}$$

## CALCULATION

$$\begin{aligned}
\text{Mass of oil } m_2 &= \ .. \ \mathrm{kg} \\
\text{Mass of ice } m &= \ .. \ \mathrm{kg}
\end{aligned}$$

Heat lost by oil and calorimeter = heat gained by ice in melting to water

If $c$ is the specific heat capacity of oil, and the specific latent heat of fusion of ice is $l$, about $3.3 \times 10^5 \ \mathrm{J\,kg^{-1}}$, then

$$(m_2 c + m_1 c_1)(t_1 - t_2) = ml + mc_w(t_2 - 0)$$

where $c_w = 4200 \ \mathrm{J\,kg^{-1}\,K^{-1}}$.

## CONCLUSION

The specific heat of the oil was .. $\mathrm{J\,kg^{-1}\,K^{-1}}$.

## ERRORS AND ORDER OF ACCURACY

See previous experiment. Verify that if the ice melts before it is placed in the calorimeter, the result for $c$ will be high.

# EXPERIMENT 21

## Investigation of: (i) Charles' Law, (ii) the Relation between Gas Volume and Absolute Temperature at Constant Pressure

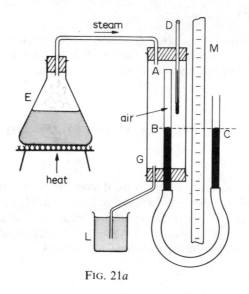

FIG. 21a

## APPARATUS

Charles' law apparatus, ABC, with metre rule M, thermometer D (0–100 °C), flask E, tripod, gauze, burner, beaker L.

## METHOD

Half-fill the flask E with water, and connect it to the steam jacket G. Connect the bottom of G to L with rubber tubing, after filling L with some water. By moving the side C of the manometer up or down, make the levels B, C of the mercury in the two sides equal, as shown; the pressure of the air in AB is then atmospheric pressure. Read the height of A and B on the metre rule M, and record also the temperature of the air in G.

Now heat the water in E so that steam passes through the jacket G. When the temperature in G is constant, record it, equate the levels in B and C again by lowering C, and then take the new reading of the level B. Turn off the burner. As the air in G cools, equate the levels B and C at three or four other temperatures, and each time note: (a) the temperature; (b) the reading of B on M.

## MEASUREMENTS

Reading of A = .. mm

| Reading of B /mm | Temp. $t$ /°C | Length AB /mm | $T$ /K (absolute) |
|---|---|---|---|
|  |  |  |  |

# (i) CHARLES' LAW

GRAPH

Plot the volume of the air, proportional to AB, v. temperature, $t$, in °C, starting the temperature-axis from 0 °C (Fig. 21b(i)). Draw the best straight line XY through the points, produce it back to cut the volume-axis at N, and record ON, the volume at 0 °C.

CALCULATION

$$\text{Cubic expansivity, } \alpha, = \frac{\text{Increase in vol. from } 0\,°C}{\text{Vol. at } 0\,°C \times \text{Temp. rise}}$$

$$= \frac{PQ - ON}{ON \times OP}$$

$$= .. \ K^{-1}$$

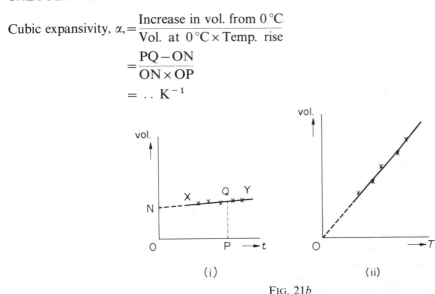

FIG. 21b

# (ii) VOLUME AND ABSOLUTE TEMPERATURE

Plot the volume of the air, AB, v. $T$ K, the absolute temperature, which is $(273 + t)$, entering the values of the latter in the table of measurements, as shown on p. 60. Start the temperature-axis from zero. See whether the points lie reasonably well on a straight line passing through the origin (Fig. 21b(ii)).

CONCLUSION

   (i) The volume of a gas increases by .. $K^{-1}$ of its volume at 0 °C.
   (ii) From the volume v. absolute temperature graph, the conclusion was ...

ERRORS

Errors occur because the temperature changes while the levels B and C are equalised; to minimise the temperature change, equalising of levels should be carried out up to the time when readings are taken. There are errors due to: (1) the readings of A and B on the metre rule; (2) the reading of the thermometer.

ORDER OF ACCURACY

To find the order of accuracy of the cubic expansivity in Charles' law, draw the lines of greatest and least slope passing through the points and recalculate the coefficient. The order of accuracy can then be found (see p. 9).

# EXPERIMENT 22

## (i) Measurement of Laboratory Temperature; and (ii) Investigation of Variation of Boiling-point of Salt Solution, using Constant-volume Gas Thermometer

### APPARATUS

Bulb A connected to mercury manometer B with metre rule C, large beaker D, common salt, chemical balance, tripod, gauze, burner, ice.

### (i) LABORATORY TEMPERATURE MEASUREMENT

#### METHOD

With the bulb A at room temperature, move G up or down until mercury level on the other side reaches a fixed mark F on the metre rule C, as shown. Note the reading of F and G, the levels of the mercury, on C. Now place the bulb A in the beaker D, and surround it com-pletely with ice. Move G up or down until the mercury level on the left side of the manometer again reaches F, so that the volume of the air in A is kept constant. Note the reading on C of the level G. Now warm the ice until it all melts, add water to cover the whole of A, and then boil the water. Raise the right-hand side of the manometer until the level on the other side reaches F again, and note the reading of G. Record the barometric pressure from a Fortin barometer, and look up the boiling point of water $X\,°C$ at this pressure.

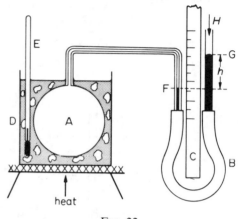

FIG. 22a

### MEASUREMENTS

Barometric pressure, $H = $ .. mmHg
Reading of F $\quad = $ .. mm

|  | Room temp. $t/°C$ | Melting ice $0\,°C$ | B.P. of water $X\,°C$ |
|---|---|---|---|
| Reading of G /m |  |  |  |
| Pressure of air $p = H + h$ /mmHg |  |  |  |

### GRAPH

Plot on a pressure $p$ v. temperature $t$ graph the pressure of the air at melting ice temperature, $0\,°C$, and at the boiling-water temperature, $X\,°C$. Suppose these are A, B respectively. Join the points by a straight line. Then read off from the line the temperature $t\,°C$ corresponding to E, the pressure of the air at the room temperature, obtained from the above table.

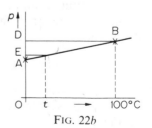

FIG. 22b

## CONCLUSION

The temperature of the room was . . °C.

## ERRORS

Errors occur: (1) in reading the metre scale; (2) because part of the air, between the bulb and the manometer, is not at the temperature of the melting ice or boiling water; (3) in setting the mercury level at the mark F.

## ORDER OF ACCURACY

The laboratory temperature $t$ is given by

$$t = \frac{OE - OA}{OD - OA} \times 100 \,°C$$

The maximum error in $(OE - OA)$ is twice the error in either measurement. So is the maximum error in $(OD - OA)$. If we denote the error in OE, OD or OA by $\Delta(OE)$,

$$\text{maximum } \% \text{ error in } t = \left[ \frac{2\,\Delta(OE)}{OE - OA} + \frac{2\,\Delta(OE)}{OD - OA} \right] \times 100\%$$

# (ii) VARIATION OF BOILING-POINT OF SALT SOLUTION WITH MASS OF SALT

## METHOD

As in the previous experiment, determine the pressure at constant volume at the temperature of melting ice ($p_0$) and at the temperature of boiling water ($p_1$). Add 10 g of common salt to the hot water in the beaker, and determine the increased pressure $p$ when the water again boils. Repeat for additional masses of 10 g of salt.

## MEASUREMENTS

Pressure at m.p. of ice, $0\,°C$ = .. mmHg
Pressure at b.p. of water, $X\,°C$ = .. mmHg

| Mass of salt /g | Pressure of air /mmHg | Temperature /°C |
|---|---|---|
|  |  |  |

## GRAPH

(*a*) Draw a pressure $p$ v. temperature $t$ straight-line graph, between 0 and $X\,°C$. From the graph, read off the boiling-points corresponding to the pressures in the above table, when salt is added to the water. (*b*) Then draw a graph of the boiling-point v. mass of salt.

## CONCLUSION

The boiling-point of water varies ... with the mass of salt added.

# EXPERIMENT 23

## Measurement of Thermal Conductivity of Good Conductor (Searle's Method)

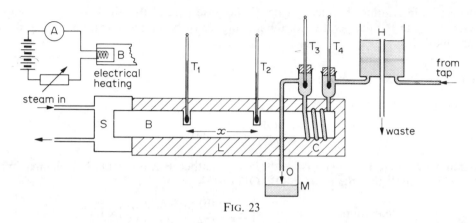

FIG. 23

### APPARATUS

Four thermometers $T_1$, $T_2$, $T_3$, $T_4$, constant-head apparatus H, measuring cylinder M, stop-watch, Searle's apparatus (see diagram), steam generator, *or* accumulators, ammeter, voltmeter, rheostat.

### METHOD

Insert the thermometers $T_1$, $T_2$ in the holes drilled in the bar B, with a little mercury or glycerine to ensure good thermal contact. Replace the lagging L carefully. Place the thermometers $T_3$, $T_4$ in the inlet and outlet tubes of the spiral C, using well-fitting rubber bungs, and connect the inlet tube to the constant-head apparatus H. Adjust the positions of the outlet tube O and H until a steady flow of a few cubic centimetres of water per second is obtained.

If the bar is supplied with a steam chamber S at the other end connect this to a steam flask and maintain a steady flow of steam. If an electrical heating coil is supplied instead, connect this in series with accumulators of suitable voltage, and an ammeter and rheostat of suitable ranges. Adjust the rheostat, if necessary, to maintain a constant current throughout the experiment.

Take readings of $T_1$ and $T_2$ every few minutes, and when they are nearly steady, adjust the height of H so that the rate of flow is increased or decreased until the difference between $T_3$ and $T_4$ is about 5 K. Allow the thermometers to reach their steady values, and read $T_1$, $T_2$, $T_3$ and $T_4$. Collect the outflow of water from C for a known time $t$ in the measuring cylinder M and record the mass $m$ of water collected. If time allows, decrease the rate of flow, and repeat the above measurements when the steady state is again reached. Measure the distance $x$ between the centres of the holes drilled in the bar, and then its diameter $d$ with calipers.

MEASUREMENTS

| | |
|---|---|
| $T_1/°C$ | |
| $T_2/°C$ | |
| $T_3/°C$ | |
| $T_4/°C$ | |
| Mass of water $m$ /kg | |
| Time $t$ /s | |
| Distance $x$ /m | |
| Diameter $d$ /m | |

CALCULATION

When the steady state is reached the heat flowing along the bar per second is equal to the heat gained per second by the water flowing in C.

$$\therefore \ k . \frac{\pi d^2}{4} \frac{(T_1 - T_2)}{x} = \frac{m}{t} c_W (T_4 - T_3)$$

where $c_W$ is the specific heat capacity of water ($4200 \ \mathrm{J \, kg^{-1} \, K^{-1}}$).
Substitute in the above equation, and calculate the thermal conductivity $k$.

CONCLUSION

The thermal conductivity $k$ of the metal was found to be .. $\mathrm{W \, m^{-1} \, K^{-1}}$.

ERRORS

1. The error in each of the temperature differences $(T_4 - T_3)$ and $(T_1 - T_2)$ is the sum of the errors in reading the two temperatures. (Loss of heat from the bar may be expected to affect $T_1$ and $T_2$, and $T_3$ and $T_4$ more or less equally, thus not affecting their differences to any great extent.)

2. Errors in measuring the mass $m$ and time $t$ can be minimised by collecting water over as long a period as possible.

3. If a good pair of calipers are used the errors in $x$ and $d$ may also be kept reasonably low compared with those mentioned above.

# EXPERIMENT 24

## Measurement of Thermal Conductivity of Bad Conductor (Lees' Method)

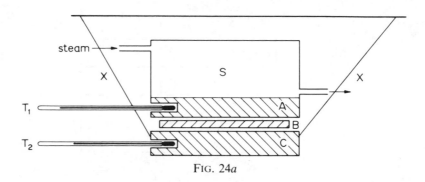

FIG. 24a

## APPARATUS

Lees' apparatus, thermometers $T_1$, $T_2$, stop-watch, steam flask, bad conductor B in the form of a disc, micrometer gauge, calipers.

## METHOD

1. Suspend the thick brass disc C by strings XX in a place away from draughts, and place the badly conducting disc B on it, using a little glycerine to ensure good thermal contact. Rest the steam chest S with its base A over B, and connect S to a steam flask. Carefully insert the thermometers $T_1$ and $T_2$ into holes drilled in A and C, and read them every few minutes until they indicate that a steady state has been reached. Record the steady readings of $T_1$ and $T_2$, $t_1$ and $t_2$ respectively.

2. Remove the steam chest, but leave B in place, and warm the disc C with a *low* bunsen flame until it is about 10 K above the steady temperature reached before. Allow C to cool, taking the temperature at frequent intervals (15–30 second) until it has fallen about 20 K.

Measure the diameter $d$ of the disc B with calipers, and its thickness $x$ with a micrometer gauge. Remove the disc $C$, and find its mass, $m$.

## MEASUREMENTS

| | |
|---|---|
| Steady temperature $t_1$/°C | |
| ,,           ,,      $t_2$/°C | |
| Diameter $d$/m | |
| Thickness $x$/m | |
| Mass of C, $m$/kg | |

| | | | | | | | | | | | | | |
|---|---|---|---|---|---|---|---|---|---|---|---|---|---|
| Temperature of $C$/°C | | | | | | | | | | | | | |
| Time /s | | | | | | | | | | | | | |

# CALCULATION

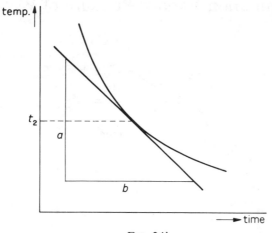

FIG. 24b

(*a*) Plot a cooling curve from the result of part (2), and draw the tangent to the curve at the temperature corresponding to the steady value of $t_2$ in part (1), as in Fig. 24b. Measure the slope $g = a/b$, which gives the rate of cooling of the disc C at temperature $t_2$. The heat lost per second in watt (W) at this temperature is:

$$m.c.g \quad . \quad . \quad . \quad . \quad . \quad . \quad . \quad . \quad . \quad . \quad . \quad . \quad (1)$$

where $m$ is in kg and $c$ is the specific heat capacity of the metal of which C is made in $J\,kg^{-1}\,K^{-1}$.

(*b*) Neglecting heat lost in part (2) by conduction through the bad conductor, the conditions are similar to those in part (1), where heat passes from A through B to C. Thus if the faces of $B$ are at $t_1°$, $t_2°\,C$,

$$m.c.g = k.\frac{\pi d^2}{4}.\frac{(t_1 - t_2)}{x}. \quad . \quad . \quad . \quad . \quad . \quad . \quad . \quad (2)$$

Hence, with $d$ and $x$ in metre, $k$ can be calculated in $W\,m^{-1}\,K^{-1}$.

## CONCLUSION

The thermal conductivity of the material was found to be .. $W\,m^{-1}\,K^{-1}$.

## ERRORS

1. Error in measuring the temperatures $t_1$ and $t_2$. Note it is assumed that the temperature is uniform throughout the brass blocks A and C.
2. Error in measuring the thickness $x$ and diameter $d$ of the disc B.
3. Error in measuring the slope $g$ of the cooling curve at temperature $t_2$.

It should be possible to measure the mass $m$ with a percentage error small compared to those mentioned above.

# EXPERIMENT 25

## Measurement of Saturation Vapour Pressure of Water (Dynamic Method)

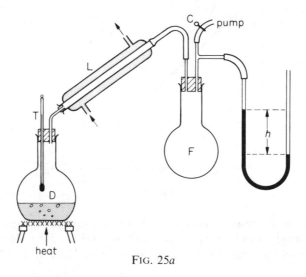

FIG. 25a

APPARATUS

Flask D, thermometer T, Liebig condenser L, large flask F, long mercury manometer, filter or vacuum pump, pressure tubing, clip C, tripod, gauze, burner.

METHOD

Place some unglazed pot and water in the flask D. Then connect D to the condenser L, and L to the large flask F. Fit a T-piece to F, join one arm of the T-piece to the manometer and the other by pressure tubing to the filter or vacuum pump. Use a clip C to close the tubing. Fit the flask D with a thermometer T, as shown. All the bungs should be of rubber, tight fitting, and greased if necessary.

Start the pump with a slow rate of pumping, and close the clip C with the manometer shows a small difference in level of the mercury. Boil the water, and read the temperature, $t$, of the vapour, and the difference, $h$, between the levels in the mercury manometer. Repeat with a faster rate of pumping, so that the difference in levels of the manometer is higher. Note the new steady difference, $h$, and the boiling-point. Repeat with faster rates of pumping, increasing the magnitude of $h$ up to the limit of the pump. Finally, read the atmospheric pressure, A, on a Fortin barometer. If necessary, adjust the clip C so that the pump just maintains a steady reading on the manometer.

MEASUREMENTS

Atmospheric pressure $A =$ .. mm.

| Temperature $t/°C$ | | | | | | | | | | |
|---|---|---|---|---|---|---|---|---|---|---|
| 'Head' $h/mmHg$ | | | | | | | | | | |
| $(A-h)/mmHg$ | | | | | | | | | | |

GRAPH

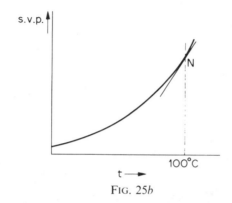

FIG. 25b

The saturation vapour pressure, $p = A - h$, where $A$ is the atmospheric pressure and $h$ is the difference in the mercury levels, or 'head', of the manometer. Calculate $p$ at each temperature and insert the values in the table of measurements.

Plot a graph of $p$ against temperature $t$ (Fig. 25b), drawing a smooth curve among the plotted points.

Draw the tangent $N$ to the graph at $t = 100\,°C$, and calculate the change in boiling-point per mmHg change in the pressure from the slope of the tangent.

CONCLUSIONS

1. The variation of the s.v.p. of water with temperature in the range .. °C to 100 °C is shown by the graph.

2. A change of 1 mmHg in atmospheric pressure produces a change of .. K in the boiling-point, in the region of 100 °C.

ERRORS

1. Error in reading the head $h$.
2. Error in reading the atmospheric pressure (this is small if a Fortin barometer is used).
3. Error in reading the thermometer.
4. Difficulty in keeping the conditions steady for each reading.

# PART FOUR

# Geometrical Optics

# EXPERIMENT 26

# Measurement of Focal Length of Concave Mirror by Locating Centre of Curvature

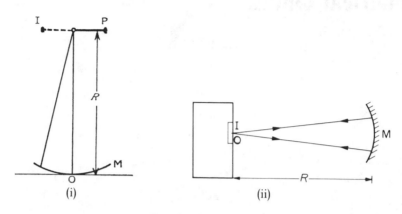

FIG. 26

## APPARATUS

Concave mirror M, metre rule, pin P, clamp, retort stand (Fig. 26(*i*)).

*Alternative.* Concave mirror M, metre rule, illuminated object O such as crosswires, holder for mirror (Fig. 26(*ii*)).

## METHOD

*Pin.* Place the concave mirror $M$ on a flat, horizontal surface. Fix the pin P in the jaws of the clamp, so that: (*a*) the point of the pin is on the axis of the mirror; (*b*) the pin is horizontal. With the eye at least 25 cm above P, look for the image I of P. If the image is erect, P is within the focal point of the mirror, and must be moved farther away, until the image is inverted. Adjust the pin P so that the point of P and its image appear to touch. Then move P until there is no parallax between it and the image, i.e. until there is no relative motion between P and I as the eye is moved from side to side. Measure the distance $R$ between the pole O of the mirror and the pin P. Repeat the setting of the no-parallax position four times, after displacing the pin, twice upwards and twice downwards. Record the distance $R$ in each case.

*Illuminated object.* Place the illuminated object O in front of the mirror M at the same height as the middle (pole) of M (Fig. 26(*ii*)). Move M until an image I is in sharp focus beside the object O. Measure $R$, the distance MO. Now find the maximum error in $R$ by moving M until the image is *just* out of focus and record the uncertainty in $R$; do this in both directions, towards and away from the object O.

Take four readings of $R$, and enter the results in the measurements.

## MEASUREMENTS

| Distance $R$ /cm | (1) .. | (2) .. | (3) .. | (4) .. | Average .. |
|---|---|---|---|---|---|

## CALCULATION

When object and image coincide in concave mirror, object distance $=R=2f$.

$$\therefore \; f=\frac{R}{2}= \; .. \; cm$$

## CONCLUSION

The focal length was found to be .. cm.

## ERRORS. ORDER OF ACCURACY

Error in (1) setting the zero of the rule at the pole of the mirror, (2) reading the pin *or* illuminated object position, (3) setting the no-parallax position for the pin *or* setting the image in sharp focus beside the illuminated object. The total error in $R$, and hence in $f$, is the sum of the errors.

# EXPERIMENT 27

## Measurement of Focal Length of Concave Mirror by Object and Image Method

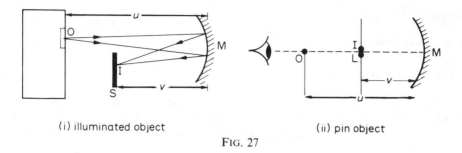

(i) illuminated object          (ii) pin object

FIG. 27

### APPARATUS

Illuminated object O, screen S, concave mirror M, metre rule or optical bench, holder for mirror.
*Alternative.* Concave mirror M, two pins in corks, stands and clamps, metre rule or optical bench.

### METHOD

*Illuminated object.* Place the illuminated object O in front of the mirror M at the same height as the centre of the mirror. Adjust the position of O so that a reduced image I is in sharp focus on the screen S when S is between O and M. Measure OM($u$) and IM($v$) to the nearest millimetre. Obtain two more different values of $u$ and $v$.

Now place the screen S farther than O from the mirror M. By moving O towards the mirror M and adjusting S, a magnified image of O can be obtained in sharp focus on S. Measure OM($u$) and IM($v$). Obtain two other values of $u$ and $v$ for a magnified image.

*Pin.* Set up a pin O in front of the mirror M so that an inverted real image I of the pin is seen (Fig. 27(i)). Adjust the top of O to be on the same level as the centre of M. Take the second pin L, place it between O and M, and move it until there is no parallax between L and the image I on moving the eye from side to side. Measure OM($u$) and LM($v$), to the nearest millimetre. Vary OM twice more, obtaining a real image each time, and measuring $u$ and $v$ on each occasion.

Now place the pin O close to M, so that an erect magnified virtual image I is obtained behind M. With a pin L *behind* the mirror, move L until I and L have no parallax between them. Measure OM($u$) and ML($v$). Obtain two more values of $u$ and $v$ for a virtual image.

### MEASUREMENTS

| Object distance $u$ | Image distance $v$ | Focal length $f$ |
|:---:|:---:|:---:|
| . . | . . | . . |

74

## CALCULATION

Calculate $f$ for each pair of values of $u$ and $v$ from

$$\frac{1}{v} + \frac{1}{u} = \frac{1}{f}$$

With the 'Real is Positive' convention, $v$ is *negative* when the image is virtual.

## GRAPHICAL METHOD

From tables, obtain the reciprocals $1/v$ and $1/u$, and enter them in the table of measurements. Then plot $1/v$ v. $1/u$, plotting the values on each axis from the zero. Draw the best straight line, produce it to intersect each axis, and read off the two intercepts, $x$ and $y$. Calculate $f$ from $1/x$ and $1/y$, and obtain the average value.

## CONCLUSION

The focal length of the concave mirror was . . cm.

## ERRORS. ORDER OF ACCURACY

*Illuminated object*. Determine the error due to the uncertainty of setting the image in sharpest focus (p. 72).

Errors are also made in the measurement of $u$ and $v$.

*Pin*. The most important error is the uncertainty in finding the 'no parallax' position—this can be estimated by noting the distance the pin has to be moved so that there is *just* parallax between it and the image.

*Order of accuracy*. From the six values of $f$, obtain the greatest difference from the average value. With the graphical method, determine the accuracy as explained on p. 9.

# EXPERIMENT 28

## Measurement of Focal Length of Convex Mirror with Converging Lens

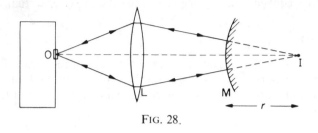

FIG. 28.

APPARATUS

Light box O, converging lens L, convex mirror M, screen, metre rule or optical bench.

METHOD

Place the converging lens L near to the illuminated object O, and adjust the screen beyond L so that a magnified image is obtained clearly on it. If necessary, move L farther away from O. Measure the distance from L to I, the image position. Note that LI must be greater than the radius of the mirror M. Now interpose the mirror M between L and I, and move M until a clear image is received back besides the object O. Measure the distance LM.

Altering the distance of L from O, repeat the experiment four times.

MEASUREMENTS

| LI | LM | $r = LI - LM$ |
|----|----|----|
|    |    |    |

Average $r = $ .. cm

CALCULATION

Since the image is formed back besides O, the rays strike M normally in Fig. 28. Hence radius of curvature, $r, = MI = LI - LM$. Calculate the average value of $r$ from the five measurements. The focal length $f = r/2$.

CONCLUSION

The focal length of the convex mirror was .. cm.

## ERRORS, ORDER OF ACCURACY

Errors occur in measuring LI and LM, and in positioning the screen and mirror for sharpest focus.
From the five values of $r$, obtain the greatest difference from the average result.

## NOTE

Repeat the experiment using a well-illuminated pin at O in place of the light box (Fig. 28). Take several measurements, work out the average value of $r$ and hence of $f$. Compare the accuracy with that obtained by using the light box.

# EXPERIMENT 29

## Measurement of Refractive Index of Transparent Block and Liquid by Apparent Depth

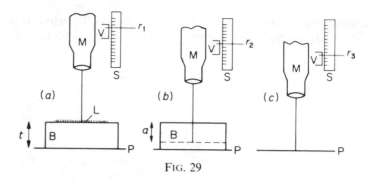

FIG. 29

APPARATUS

Glass or perspex block B, travelling microscope M, lycopodium powder L, beaker.

METHOD

1. *Glass.* Place the block B on a sheet of paper P, and arrange the travelling microscope so that the microscope M and the scale S are vertical. Focus the microscope M on the upper surface of the block, using a little lycopodium powder L or chalk dust if necessary (Fig. 29 (a)). Having achieved a sharp focus using the fine adjustment screw, take the reading $r_1$ of the vernier V on the scale S.

Move the microscope down until the paper, seen through the block, is in sharp focus. Take the reading $r_2$ of the vernier (Fig. 29 (b)). Remove the block, and focus on the paper, and take the vernier reading $r_3$ (Fig. 29 (c)). Turn the block onto another face and repeat the measurements.

2. *Water.* Repeat the experiment for water by focusing on a pin (or grains of sand) on the bottom of a beaker, then partly filling the beaker with water and re-focusing, and then sprinkling a little lycopodium powder on the water and focusing on the top of the water.

MEASUREMENTS

| $r_1$/mm | $r_2$/mm | $r_3$/mm | $(r_1 - r_2)$/mm | $(r_1 - r_3)$/mm |
|---|---|---|---|---|
| | | | | |

CALCULATION

The refractive index $n$ is given by:

$$n = \frac{\text{Real thickness of block}}{\text{Apparent thickness of block}} = \frac{t}{a} = \frac{r_1 - r_3}{r_1 - r_2} = \quad .. \text{ Expt. (1)}$$

$$= \quad .. \text{ Expt. (2)}$$

$$= \quad .. \text{ Average}$$

## CONCLUSION

The refractive index of the block was found to be ...; the refractive index of the water was found to be ...

## ERRORS. ORDER OF ACCURACY

1. Errors in focusing the microscope may be investigated by reading the distance travelled by the microscope when it is moved from a position just out of focus too close to the object, to one just too far from it. 2. Error in reading the vernier.

The maximum error in $t$, or $(r_1 - r_3)$ is $2\,\Delta r$, where $r$ is the error in $r_1$ or $r_3$; likewise, the maximum error in $a$, or $(r_1 - r_2)$, is $2\,\Delta r$. So

$$\text{maximum \% error in } n \ (t/a) = \left( \frac{2\,\Delta r}{t} + \frac{2\,\Delta r}{a} \right) \times 100\%$$

# EXPERIMENT 30

# Measurement of the Refractive Index of a Liquid using an Air Cell

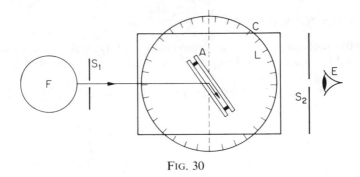

FIG. 30

APPARATUS

Rectangular glass or perspex container C, air cell A, fitted with a scale of degrees, slits $S_1$, $S_2$, sodium flame F, water and other transparent liquids.

METHOD

Fill the cell C with water, and set up the slits $S_1$, $S_2$ perpendicular to a pair of opposite faces of C, as in Fig. 30 above. Place a sodium flame F so that it can be seen through $S_1$, $S_2$ by the eye at E. Immerse the air cell A with its plane vertical, and rotate the cell slowly until the light through $S_2$ is just cut off. Read the angle $\theta_1$ on the scale attached to A. Rotate the cell to the position on the other side of the line $S_1 S_2$ at which the cut-off again occurs, and read the angle $\theta_2$ on the scale. Repeat the measurements with other liquids, as time allows.

MEASUREMENTS

|  | $\theta_1$ | $\theta_2$ | $\dfrac{\theta_1 - \theta_2}{2}$ |
|---|---|---|---|
| 1st liquid |  |  |  |
| 2nd liquid |  |  |  |
| 3rd liquid |  |  |  |

CALCULATION

Enter the values of the critical angle $c = \dfrac{\theta_1 - \theta_2}{2}$ in the table of measurements.

Calculate the values of $n$ from the formula

$$n = 1/\sin c = \ \ldots$$

80

# 30. (*Continued*)

CONCLUSIONS

1. The refractive index of .. was found to be ...
2. „      „      „      „ ·· „      „      „ „ ···
3. „      „      „      „ ·· „      „      „ „ ···

ERRORS

The errors in setting and reading the angle at which the cut-off occurs may be investigated by:

(*a*) repeating each setting of the air cell several times;
(*b*) measuring how far the cell has to be rotated near the cut-off point to change from maximum to minimum brightness.

ORDER OF ACCURACY

The error in the critical angle is equal to the error in $\theta_2 - \theta_1$. The error in $n$ is equal to the change in $1/\sin c$ produced by a change equal to the error in $c$, which may be found by inspection of sine and reciprocal tables. For example, if the estimated error in $c$ is $0.5°$, with $c = 40°$.,

$$1/\sin 40° = 1.556$$
$$1/\sin 40.5° = 1.540$$
$$\text{difference} = 0.016$$
$$\therefore \text{ error in } n = 0.016 = 0.02 \text{ approx.}$$
$$\therefore n = 1.56 \pm 0.02$$

# EXPERIMENT 31

# Measurement of the Refractive Index of Glass using a Spectrometer

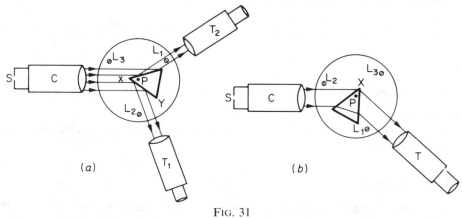

(a)  (b)

FIG. 31

APPARATUS

Prism spectrometer, glass prism P, sodium burner.

METHOD

### 1. Adjustment of Spectrometer

(a) Adjust the eyepiece of the telescope T so that the crosswires are sharply focused.

(b) Focus the telescope for parallel light using a distant object. There should be no parallax between the image seen in the telescope and the crosswires seen through the eyepiece.

(c) Place the sodium flame near the slit, which may be fairly widely open at this stage, and with the telescope and collimator in line, adjust the *collimator* so that the slit is sharply focused in the telescope. Check that there is no parallax between the images of the slit and the crosswires.

(d) *Levelling the prism.* Place the prism P on the table with one apex near the axis of the table, and turn it so that one face $(XY)$ is perpendicular to the line joining two of the levelling screws $(L_1, L_2)$ as in Fig. 31 (a). Turn the table until light can be seen (with the eye) reflected from the two faces of the prism, as in Fig. 31 (a). Set the telescope at $T_1$ so that the slit is seen through it, light being reflected from face XY. Adjust the levelling screw $L_1$ or $L_2$ until the image of the slit is central in the field of view. Rotate the telescope to $T_2$, and adjust $L_3$ until the image is central in this position. Repeat, until the image is central whichever face is viewed.

### 2. Measurement of Prism Angle

Make the slit as narrow as is practicable, and carefully set the image of the slit at $T_1$ and $T_2$ on the crosswires, reading the angles $\theta_1$ and $\theta_2$ on the vernier attached to the telescope.

### 3. Measurement of the Angle of Minimum Deviation

Rotate the table, without touching the prism, until light is refracted through the sides containing the apex X, as in Fig. 31 (b). Locate the image of the slit after refraction and rotate the table, keeping the image in view in the telescope T, until the position of minimum deviation is found. At this point the image will stop and reverse its direction of motion as the table is turned. Set the table so that the slit is in the stationary (minimum deviation) position, and use the slow-motion screw fitted to the telescope to set the crosswires accurately on the centre of the slit. Read the angle $\theta_D$ on the vernier scale. Turn the table so that light is refracted on the other side of the axis of the collimator, and again locate the position of minimum deviation on this side. Read the angle $\theta_D'$ on the vernier.

## MEASUREMENTS

| Prism Angle $A$ | Minimum Deviation $D$ |
|---|---|

$$\theta_1 = \overset{\circ}{..}$$
$$\theta_2 = \overset{\circ}{..}$$
$$A = \frac{\theta_1 - \theta_2}{2} = \overset{\circ}{..}$$

$$\theta_D = \overset{\circ}{..}$$
$$\theta_D' = \overset{\circ}{..}$$
$$D = \frac{\theta_D - \theta_D'}{2} = \overset{\circ}{..}$$

## CALCULATION

Calculate the angle of the prism $A \qquad = (\theta_1 - \theta_2)/2 = \overset{\circ}{..}$

" " " " minimum deviation $D = (\theta_D - \theta_D')/2 = \overset{\circ}{..}$

Calculate the refractive index $n$ of the glass from:

$$n = \frac{\sin \frac{1}{2}(A + D)}{\sin \frac{1}{2}A} = \ldots$$

## CONCLUSIONS

The refractive index of the glass in sodium light was found to be ...

## ERRORS

In each value of an angle $\theta$, there are errors in:

1. Setting the crosswires on the image of the slit. 2. Reading the vernier.

The error (1) may be estimated by moving the telescope until the crosswire is just off-centre, and re-reading the vernier.

3. There may also be error in setting the prism at minimum deviation, but this should have little effect, as a small prism movement near minimum deviation produces a negligible variation in $D$.

## ORDER OF ACCURACY

The errors $\delta A$ and $\delta D$ in $A$ and $D$ are given by half the sum of the errors in the angles $\theta_1$, $\theta_2$ and $\theta_D$, $\theta_D'$ respectively.

To find the maximum error in $n$, (a) increase $A$ and $D$ by $\delta A$ and $\delta D$ in the numerator of the formula for $n$, (b) decrease $A$ by $\delta A$ in the denominator, and then calculate the new value of $n$. The difference between this value and that calculated without taking account of errors gives the maximum error in $n$.

## ADDITIONAL

Draw the outline of a $60°$ prism on a sheet of white paper. From a point O in the middle of one side, draw lines such as OP which make angles $i$ of $20°$, $30°$, $40°$, ... with the normal at O (Fig. 31c). Using a ray-box (or pins), track the path of rays through the prism which are incident at O along the lines drawn. POQR is a typical ray passing through the prism.

Measure the deviation $d$ for each angle of incidence $i$. Plot a graph of $d$ against $i$. From your graph, state (a) the minimum deviation and the corresponding angle of incidence, (b) the maximum deviation and a corresponding angle of incidence. In (b), is there more than one angle of incidence for which the deviation is a maximum?

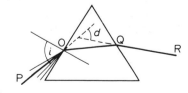

FIG. 31c

# EXPERIMENT 32

## Measurement of Focal Length of Converging Lens by Plane Mirror

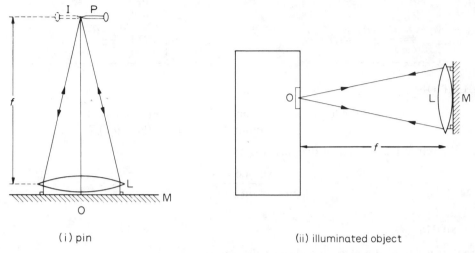

(i) pin                    (ii) illuminated object

FIG. 32

## APPARATUS

Thin plane mirror M, converging lens L, pin P, metre rule, stand and clamp.

*Alternative.* Illuminated object such as crosswires, thin plane mirror M, converging lens L, metre rule, lens holder.

## METHOD

*Pin.* Place the lens L on top of the plane mirror M, which rests on a horizontal surface (Fig. 32 (*i*)). Fix the pin P in a clamp so that: (*a*) it is horizontal, and (*b*) the point of the pin is vertically above the centre O of the lens. With the eye at least 25 cm above P, look for the inverted image I. If the image is magnified P is too close to the lens; if it is diminished P is too far from the lens. Move P up or down until there is no parallax between the point of the pin P and that of its image I, as shown in Fig. 32. Measure the distance $f$ from the centre of the lens to the pin P.

Repeat the setting of the no-parallax position, twice having previously moved P downwards out of adjustment, and twice upwards.

*Illuminated object.* Arrange the illuminated object O to be at the same height as the middle of the lens L (Fig. 32 (*ii*)). Place the plane mirror M behind L so as to reflect back the light from O passing through the lens.

Starting from some distance away from the lens, move O towards L until an image I appears in sharp focus beside O. If necessary, tilt the mirror. Measure the distance $f$ from the centre of the lens to O. Determine the uncertainty in focusing the image I by moving O in either direction towards, or away from, the lens L until the image is *just* out of focus.

Repeat for three more settings when the image is in sharp focus beside O, one by moving O towards the lens and two by moving O away from it.

## MEASUREMENTS

| Distance $f$ from centre of lens to pin /cm | (1) | (2) | (3) | (4) | Average |
|---|---|---|---|---|---|
| | | | | | |

## CALCULATION

The distance from the pin P, or object O, to the centre of the lens when P or O coincides with its image I = focal length $f$ = .. cm.

## CONCLUSION

The focal length $f$ = .. cm.

## ERRORS

1. Error in setting the zero of the scale at the centre of the lens.
2. Error in reading the position of P.
3. Error in locating the position of no parallax. The size of this error may be estimated from the variations in the measurements of $f$.

## ORDER OF ACCURACY

The error, $\delta f$, in $f$ = sum of errors due to (1), (2) and (3) above. Thus the order of accuracy can be obtained.

## NOTE

The *power P* of a lens is given by $P = 1/f$. Lenses with short focal length thus have high power. The unit of $P$ is the 'dioptre', symbol, D. A converging lens of $f = 1$ metre has a power of $+1$ D. A diverging lens with $f = 10\,cm = 0.1$ metre has a power of $-10$ D. From your result for $f$, calculate the power of the lens used.

## ADDITIONAL

### *Diverging Lens Focal Length*

Obtain a *diverging* lens of smaller power than the converging lens used, so that the two lenses in contact together act as a converging lens.

Using the plane mirror as described, measure the combined focal length $F$ of the two lenses *in contact*—take the average of three measurements. Deduce the focal length of the diverging lens from the relation

$$\frac{1}{F} = \frac{1}{f_1} + \frac{1}{f_2}$$

*Question.* Do you consider this is an accurate method of measuring the focal length of a diverging lens? Give a reason for your answer.

# EXPERIMENT 33

## Measurement of Focal Length of an Inaccessible Converging Lens by Displacement Method

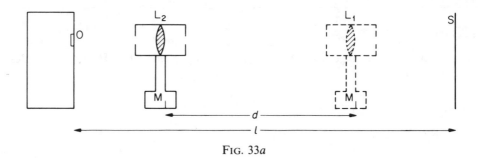

FIG. 33a

APPARATUS

Illuminated object O, screen S, converging lens mounted in a short tube on a stand carrying a mark M, metre rule.

METHOD

Measure the focal length of the lens approximately by focusing the image of a distant object, a window, for example, on the wall. Set the illuminated object O and screen S about six times the focal length apart, and measure their separation $l$ carefully. Move the lens about near the screen until a *diminished* image is formed on the screen with the lens at $L_1$, and adjust its position until the image is sharply focused. Record the reading of the mark M on the lens holder base on a metre rule laid along the bench. Move the lens away from the screen until, at a second position $L_2$, a *magnified* image is formed on the screen. Again adjust the position of the lens for the sharpest focus, and record the new reading of the mark M. The difference between the two readings is the displacement $d$ of the lens.

Vary $l$ over as wide a range as possible, and repeat the measurement of $d$, taking about six sets of readings.

MEASUREMENTS

| 1st reading of mark $M$ /cm | 2nd reading of mark $M$ /cm | Displacement $d$ /cm | Separation $l$ /cm | $l^2 - d^2$ /cm$^2$ |
|---|---|---|---|---|
|  |  |  |  |  |

CALCULATION

By the usual theory,
$$f = \frac{l^2 - d^2}{4l}$$

hence
$$l^2 - d^2 = 4f.l \quad . \quad . \quad . \quad . \quad . \quad . \quad . \quad . \quad . \quad . \quad . \quad (1)$$

86

## GRAPH

Plot a graph of $(l^2 - d^2)$ v. $l$ (Fig. 33b).

By equation (1) the gradient of the line $= a/b = 4f = $ .. cm.

$$\therefore f = \text{.. cm.}$$

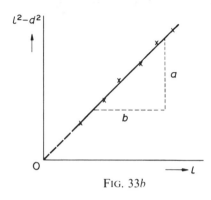

FIG. 33b

## CONCLUSION

The focal length of the lens was .. cm.

## ERRORS

1. 'Zero' error and reading error in $l$.
2. Errors in reading the positions of the mark on the lens holder.
3. Errors in setting the image at its sharpest focus. Find how far the lens may be moved in either direction without perceptibly affecting the focusing; this may be quite large.

## ORDER OF ACCURACY

Draw the lines of greater and lesser slope, passing through the origin, which just fail to agree with a majority of the plotted points. Find the variation in the values of $f$ obtained from their slopes, and hence the order of accuracy of $f$.

## ADDITIONAL

### *Minimum Distance between Object and Image* for *Converging Lens*

From your values of $u$ and $v$, plot the value of $(u+v)$ against the corresponding value of $u$. If necessary, obtain additional values of $u$ and $v$ which help you to draw the minimum of the graph more accurately.

From your graph, obtain the minimum value of $(u+v)$ and the corresponding value of $u$. Calculate the ratio of each of these values to $f$, the focal length of the lens.

# EXPERIMENT 34

## Measurement of Focal Length of Diverging Lens using a Converging Lens

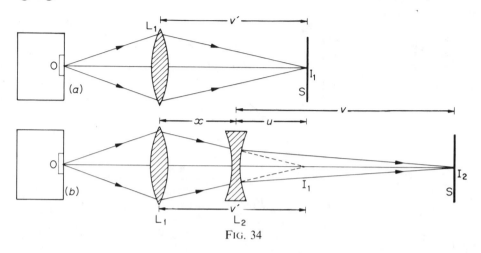

FIG. 34

APPARATUS

Illuminated object O, converging lens $L_1$, diverging lens $L_2$, screen S, metre rule.

METHOD

Set up the object O, converging lens $L_1$, and screen S, so that a sharply focused image $I_1$ is formed on the screen S, which should be about 50 cm from the lens. Measure the distance $v'$ from $L_1$ to the image $I_1$ (Fig. 34 (a)).

Place the diverging lens $L_2$ between $L_1$ and the screen, about 5 cm from the screen, and move the screen until, at $I_2$, the image is again sharply focused. Measure the distance $x$ between the two lenses, and the distance $v$ from the *diverging* lens $L_2$ to the screen at $I_2$ (Fig. 34 (b)).

Repeat the measurements for, say, four more values of $x$ and $v$, having moved $L_2$ nearer to $L_1$. Note that $L_2$ and $I_1$ will be separated by a distance equal to the focal length of $L_2$ when it becomes impossible to obtain a focused image.

MEASUREMENTS

|  | $L_1I_1=v'$/cm | $L_1L_2=x$/cm | $L_2I_2=v$/cm | $u=v'-x$/cm | $f$/cm |
|---|---|---|---|---|---|
| (1) |  |  |  |  |  |
| (2) |  |  |  |  |  |
| (3) |  |  |  |  |  |
| (4) |  |  |  |  |  |
| (5) |  |  |  |  |  |

Average value of $f=$ .. cm

88

## CALCULATION

For the diverging lens, with the 'Real is Positive' convention,

$$\frac{1}{f} = \frac{1}{-(v'-x)} + \frac{1}{v} \quad \text{(virtual object, real image)}$$

## GRAPHICAL METHOD

Obtain the reciprocals $1/v$ and $1/u$, and enter them in the table of measurements. Then plot $1/v$ v. $1/u$, plotting from the zero on each axis, draw the best straight line through the points, and measure the intercepts, $x$ and $y$, on the respective axes. Calculate $f$ from $1/x$ and $1/y$, and obtain the average value of $f$.

## CONCLUSION

The focal length of the diverging lens was .. cm.

## ERRORS

Each measurement of length is subject to: (1) error in setting the zero of the scale at the required point; (2) error in reading the scale.

## ORDER OF ACCURACY

The maximum difference between the average value of $f$ and the individual values obtained for $f$ is a fair measure of the order of accuracy. The accuracy of the graphical method can be found as explained on p. 9.

# EXPERIMENT 35

## Measurement of Focal Length of a Diverging Lens using a Concave Mirror

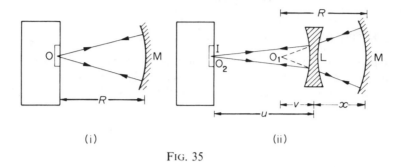

(i)                    (ii)

FIG. 35

APPARATUS

Concave mirror M, diverging lens L, illuminated object O such as crosswires, metre rule (Fig. 35).

METHOD

1. Set up the illuminated object O in front of the mirror M at the same height as the middle of the mirror (Fig. 35 (i)). Move M until the image of O is sharply in focus beside O. Measure the distance MO—this is the radius of curvature $R$ of M. Repeat this determination of $R$.

2. Now place the diverging lens L between O and M (Fig. 35 (ii)). Move the object back from a position $O_1$ to a position $O_2$ farther from the mirror, where a clear image of the object is now focused *beside the object*. This image is produced by reflection from M as shown—check this is the case by intercepting the light between L and M with a card, when the image should disappear.

Measure the distance $LO_2$ or $u$, and the distance $x$ from L to M. By moving the lens, repeat the experiment three times more and record the three sets of results.

MEASUREMENTS

| $MO_1 = R$/cm | $ML = x$/cm | $LO_2 = u$/cm | $v = R - x$/cm | $f$/cm |
|---|---|---|---|---|
| | | | | |
| | | | | |

CALCULATION

For the diverging lens, with the 'Real is Positive' convention,

$$\frac{1}{f} = \frac{1}{v} + \frac{1}{u} = \frac{1}{-(R-x)} + \frac{1}{u} \text{ (real object, virtual image)}$$

Calculate the average value for $f$ from all your results.

CONCLUSIONS

The focal length of the diverging lens was .. cm.

## ERRORS

Each measurement of length is subject to: (1) error in setting the zero of the scale; (2) error in reading the scale.

## ORDER OF ACCURACY

The maximum difference between the average value of $f$ and the individual values obtained for $f$ is a fair measure of the order of accuracy.

# EXPERIMENT 36

## Measurement of Radii of Curvature of Converging Lens (Boys' Method), and its Refractive Index

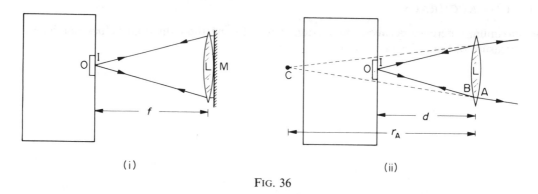

(i)                                                           (ii)

FIG. 36

## APPARATUS

Illuminated object O, converging lens L, plane mirror M, lens holder, metre rule.

## METHOD

1. Set the illuminated object O at the same height as the middle of the lens L (Fig. 36(i)). Place the plane mirror M behind L so as to reflect back the light refracted through the lens. Move O from some distance away towards L until an image is in sharp focus beside O. Measure the distance $f$ from O to the centre of the lens. Repeat twice more, displacing O each time before re-measuring the observed distance $f$.

2. Remove the plane mirror. Move the lens L towards O until an image is observed in sharp focus beside O—this image is due to *reflection at the back surface* A of the lens (Fig. 36(ii)). Measure the distance $d$ from O to the centre of the lens. Repeat for two more settings of the lens and record the results.

3. Now turn the lens round so that the other face B is now at the back in place of A. Obtain three measurements of $d$ for this face as before.

## MEASUREMENTS

|     |         |  |  |  | Average |
|-----|---------|--|--|--|---------|
| (a) | $f$/cm  |  |  |  |         |

| (b) | | Face A | | | Average | | | Face B | | | Average |
|-----|--------|--|--|--|---------|--------|--|--|--|--|---------|
|     | $d$/cm |  |  |  |         | $d$/cm |  |  |  |  |         |

92

## CALCULATION

Since $C$ is a virtual object for *refraction in the lens*, then, from $1/v + 1/u = 1/f$ *(Real is Positive)*,

$$-\frac{1}{r_A} + \frac{1}{d} = \frac{1}{f} . \qquad \dots \dots \dots \dots \dots \quad (1)$$

Substitute the values of $d$ and $f$, and calculate $r_A$:

$$r_A = \ \text{.. cm}$$

Similarly,
$$r_B = \ \text{.. cm}$$

Now substitute $r_A$, $r_B$, $f$ in
$$\frac{1}{f} = (n - 1)\left(\frac{1}{r_A} + \frac{1}{r_B}\right) \qquad \dots \dots \dots \dots \dots \quad (2)$$

$$\therefore \ n = \dots$$

## CONCLUSIONS

The radii of curvature were found to be .. cm $(A)$.

.. cm $(B)$.

The refractive index $n$ was found to be ...

## ERRORS. ORDER OF ACCURACY

1. Error in setting the image in sharpest focus. See p. 72.
2. Error in measuring the distances from O to the lens. Care should be taken to measure to the middle of the lens.

From the values of $r_A$ and $r_B$ which have maximum uncertainty, and the value of $f$ which has maximum uncertainty, use equation (2) to calculate the value of $n$ with maximum uncertainty. The maximum error in $n$ can then be obtained.

## NOTE

If a pin is used as an object, $(a)$ it should be well illuminated, $(b)$ the inverted image due to reflection at the back surface can be clearly seen when the lens is placed on black paper or, better, floated on mercury in a shallow dish.

# EXPERIMENT 37

# Measurement of Refractive Index of a Liquid using a Converging Lens and Plane Mirror

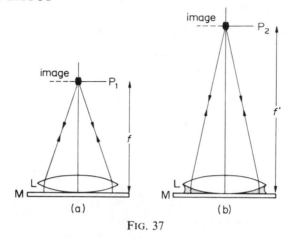

FIG. 37

## APPARATUS

Converging lens L, plane mirror M, pin, clamp and stand, liquid such as water or other transparent liquid, metre rule. Illuminated object such as crosswires.

## METHOD

1. Place the mirror M and lens L on a horizontal surface (Fig. 37 (a)). Fix the pin $P_1$ horizontally in a clamp so that its tip is vertically above the centre of the lens L. With the eye at least 25 cm above the pin, adjust it until, at $P_1$, it has no parallax with its inverted image. Repeat the setting of the no-parallax position twice more, and each time measure the distance $f$ from $P_1$ to L.

2. Moisten the mirror with a little of the liquid whose refractive index is to be found, so that the space between L and M is filled (Fig. 37 (b)). Find the new position, $P_2$, of no parallax, and make three settings of this, as above, measuring the distance $f'$ from $P_2$ to M in each case.

3. Measure the radius of curvature of the *lower* face of the lens with the illuminated crosswires by Boys' method (see p. 92). In this case, find the distance $d$ of the lens to the object when an image due to reflection at the *back* surface of the lens is in sharp focus beside the object. Repeat the measurement of $d$.

## MEASUREMENTS

|  |  |  |  | Average |
|---|---|---|---|---|
| $f = P_1 M$ /cm |  |  |  |  |
| $f' = P_2 M$ /cm |  |  |  |  |

RADIUS OF CURVATURE (see p. 92 and Fig. 36)

Object distance $d =$ .. cm, .. cm
Average $\qquad d =$ .. cm

## CALCULATION

1. If $f$ is the focal length of the converging lens, found in the first experiment, $f_l$ is the focal length of the liquid lens, and $f'$ is the focal length of the combined liquid and glass lens, then

$$\frac{1}{f_l} = \frac{1}{f'} - \frac{1}{f}$$

$$\therefore f_l = \ldots \text{ cm} \quad . \quad . \quad . \quad . \quad . \quad . \quad . \quad . \quad . \quad (a)$$

2. Radius of curvature (see p. 93).

From

$$-\frac{1}{r} + \frac{1}{d} = \frac{1}{f}$$

$$\therefore r = \ldots \text{ cm} \quad . \quad . \quad . \quad . \quad . \quad . \quad . \quad . \quad . \quad (b)$$

3. Also, $\dfrac{1}{f_l} = (n_l - 1)\dfrac{1}{r}$, since the lower face of the liquid lens is plane

$$\therefore n_l = 1 + \frac{r}{f_l} = \ldots \quad . \quad . \quad . \quad . \quad . \quad . \quad . \quad (c)$$

## CONCLUSION

The refractive index of the liquid was found to be ...

## ERRORS

1. Error in (i) setting the no-parallax position—this may be found by moving the pin until parallax is just observable, (ii) focusing the image in determining the radius of curvature—find the uncertainty as on p. 72.

2. Error in measuring $f$ and $f'$ and $r$.

## ORDER OF ACCURACY

From the formulae for $(n_l - 1)$, the maximum percentage error in $(n_l - 1)$, which is the maximum percentage error in $n_l$, is the sum of the maximum percentage errors in $f_l$ and $r$. Determine the maximum percentage error in $f_l$ from $(a)$ and in $r$ from $(b)$.

# EXPERIMENT 38

# Angular Magnification of an Astronomical Telescope

### APPARATUS

Four converging lenses, focal lengths 5, 10, 25 and 50 cm respectively; metre rule, plasticene (or lens holders), mains pearl bulb, two thin small plane mirrors, screen with millimetre scale pasted on it.

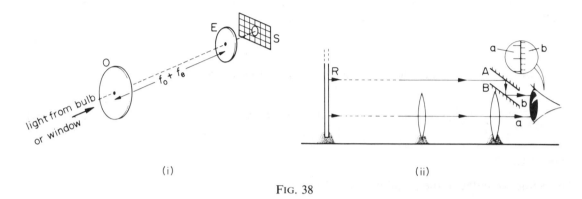

(i)                                                    (ii)

FIG. 38

### METHOD

1. Using a distant object such as a window, and measuring the distance from the lens to the image, identify the four lenses from this estimation of their focal lengths.

2. Set up a simple astronomical telescope using the 25 cm lens as the objective lens O and a 10 cm lens as the eyepiece E (Fig. 38 (i)). The distance between the lenses O and E should be equal to the sum of their focal lengths. The lenses can be held upright on the ruler using plasticene.

3. *Eyering.* (a) Place a pearl bulb in front of the objective O to illuminate this lens (Fig. 38 (i)). Move the graduated screen S slowly away from the eyepiece E until the small circular image of O *formed by the eyepiece* is sharply in focus on S. This is the eyering. (Do NOT look through the eyepiece at the pearl bulb.)

Measure and record (i) the distance of the eyering from E, (ii) the diameter $d_e$ of the eyering.

(b) Now move the telescope near the window so that the light falls directly on the objective O, move the screen S until the eyering is in focus, and repeat the measurements (i) and (ii) above.

(c) Remove the screen. Observe the field of view through the window looking through the telescope with your eye close to the eyepiece E and then moving slowly back. Is the field of view greatest at the eyering?

4. *Angular magnification.* Clamp a metre rule R vertically in a well-lit place, several metres away from the telescope, as shown in Fig. 38 (ii). Place the eyes so that R can be seen with the right eye through the telescope and by the left eye viewed directly—a simple method is to use a mirror periscope, A, B, as shown.

Adjust the telescope and periscope until say 100 mm seen directly on R can be compared with a number of magnified divisions of the image seen through the telescope. See inset in Fig. 38 (ii). Record the results.

*Alternatively,* obtain an estimate of the angular magnification by looking through the telescope at some distant window panes or bricks and comparing the width of those seen with one pane or brick viewed directly.

5. Set up another simple astronomical telescope, this time using the 50 cm focal length lens as the

96

objective and the 10 cm focal length lens as the eyepiece. The distance between the lenses should be the sum of the focal lengths. Repeat the measurements (i) and (ii) in connection with the eyering and the measurement of the angular magnification.

6. If time permits, repeat with a simple astronomical telescope using a 25 cm focal length lens as the objective and a 5 cm lens as the eyepiece.

MEASUREMENTS

|  | 25 cm & 10 cm | 50 cm & 10 cm |
|---|---|---|
| Distance of eyering from eyepiece |  |  |
| Diameter of eyering, $d_e$ |  |  |
| Angular magnification |  |  |

CALCULATIONS

In Optics theory, the following formulae are obtained for a simple astronomical telescope in normal adjustment with an objective of focal length $f_0$ and diameter or aperture $d_0$, and an eyepiece of focal length $f_e$:

(a) Distance of eyering from eyepiece $= (f_0 + f_e)(f_e/f_0)$
(b) Diameter of eyering $= (f_e/f_0) \times d_0$
(c) Angular magnification $= f_0/f_e$

Calculate each of these values for the telescopes you have made, and compare them with your measured values. Comment on the agreement between theory and your measurements and give reasons for any wide disagreement.

QUESTIONS

1. When changing from an objective of 25 cm focal length to one of 50 cm focal length, keeping the eyepiece of 10 cm focal length, what is the effect on the size of the eyering and the angular magnification?

2. When changing from a 10 cm focal length eyepiece to one of 5 cm focal length, keeping the objective of 25 cm focal length, what is the effect on the size of the eyering and the angular magnification?

3. Which telescope would you choose for looking at the planets and why?

# EXPERIMENT 39

## Angular Magnification of a Simple Microscope (Magnifying Glass)

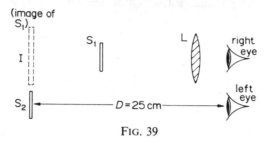

FIG. 39

APPARATUS

Converging lens L, half-metre scales $S_1$, $S_2$, metre rule, plane mirror, pin.

METHOD

Measure the focal length $f$ of the lens L by a plane-mirror method (p. 84). Then clamp the rule $S_1$ vertically, and place the lens L at a distance from it a little less than the focal length of L, so that the magnified erect image I of $S_1$ can be seen with one eye close to L. Clamp the scale $S_2$ vertically beyond $S_1$ as shown, and place it at the closest distance of distinct vision $D$. (Spectacles, if normally worn, should be retained throughout the experiment.) Move $S_2$ until it can be seen by the other eye (see Fig. 39).

Observe the length on the image of $S_1$ which appears to be equal to a given length, say 10 cm, on $S_2$ seen directly.

MEASUREMENTS

Length on $S_2$ = .. cm
Corresponding length on $S_1$ = .. cm
Least distance of distinct vision $D$ = .. cm
Focal length $f$ = .. cm

CALCULATION

If the image I is distance $D$ from the eye the theoretical magnifying power is given by

$$M = \left( \frac{D}{f} + 1 \right)$$

where $D$, $f$ are numerical values.

Calculate the theoretical magnifying power = ...

Calculate the actual magnifying power $= \dfrac{\text{Length on } S_2}{\text{Corresponding length on } S_1} = $ ...

CONCLUSION

The values of the magnifying power were:
(a) theoretical ...          (b) actual ...

# EXPERIMENT 40

# Angular Magnification of a Compound Microscope

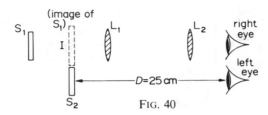

FIG. 40

## APPARATUS

Two short focus converging lenses $L_1$, $L_2$ and holders; two half-metre rules $S_1$, $S_2$, plane mirror, pin.

## METHOD

Measure the focal lengths $f_1$ and $f_2$ of $L_1$ and $L_2$ by a plane-mirror method (p. 84). Clamp the half-metre rule $S_1$ vertically, and place the lens $L_1$ a little more than its focal length distant from $S_1$. Move $L_2$ in front of $L_1$ until a magnified inverted image I of $S_1$ can be easily focused with the eye near $L_2$. Clamp $S_2$ vertically at the least distance of distinct vision $D$ from the eyes, and adjust its position until both it and the image I of $S_1$ can be simultaneously seen in focus, and edge to edge (see Fig. 40). Observe the length on the image I that corresponds to a length of, say, 10 cm on $S_2$. Record the distance $u$ cm from $S_1$ to $L_1$.

## MEASUREMENTS

| | |
|---|---|
| Length on $S_2$ | = . . cm |
| Corresponding length on $S_1$ | = . . cm |
| Object distance $u = S_1 L_1$ | = . . cm |
| Least distance $D$ of distinct vision | = . . cm |
| Focal length $f_1$ of $L_1$ | = . . cm |
| „     „   $f_2$ of $L_2$ | = . . cm |

## CALCULATION

1. Calculate the magnifying power $M$ from the relation:

$$M = \frac{\text{Length on } S_2}{\text{Corresponding length on } S_1}$$

2. Substitute the value of $u = S_1 L_1$ and $f_1$ in the lens formula, and calculate the distance $v$ of the intermediate real image from $L_1$. Substitute the values of $v, f_1, f_2,$ and $D$ in the formula for $M$:

$$M = \left(\frac{v}{f_1} - 1\right)\left(\frac{D}{f_2} + 1\right)$$

Compare the measured value of the magnifying power with that obtained from the theoretical formula. Account for any discrepancy not due to errors of measurement.

## ERRORS

1. The error in comparing the two scales may be estimated by having several observers take independent readings.

2. Errors occur in $D, f_1, f_2,$ and $u$.

The formula assumes: (a) The eye is close to the lens $L_2$. (b) The final image is at the near point. Discuss the validity of these assumptions in your experiment.

# EXPERIMENT 41

# Investigation of Photo-electric Cell and Inverse Square Law

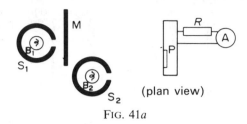

(plan view)

FIG. 41*a*

## APPARATUS

Photo-electric cell P (the copper oxide barrier layer type is convenient); microammeter A; resistance box $R$, $0$–$10\,000\,\Omega$; two motor headlamp bulbs $B_1$, $B_2$ in suitable holders with power supply; metre scale or optical bench on which a bulb and the cell can be mounted; means of darkening the part of the laboratory used; blackened masking screen M; two cylindrical opaque shields $S_1$, $S_2$, which fit over the bulbs with small apertures cut out opposite the bulb filaments and which are painted matt black inside and out (these may be cut from cardboard 'posting' tubes).

## 1. LINEARITY OF THE CELL AND METER COMBINATION

### METHOD

Place the photocell facing the two bulbs $B_1$, $B_2$. Connect the photocell P, the resistance box $R$ and the meter A in series as in Fig. 41*a*, and arrange that as little stray light as possible reaches the cell. With one bulb switched on, vary $R$ until the meter is fully deflected when the bulb is about $20\,cm$ from P. Record the deflection of the meter, if any, when both bulbs are switched off, and take all subsequent deflections from this value as zero.

Switch on both bulbs, and place the shields $S_1$, $S_2$ over them. Turn the aperture of $S_1$ so that light from $B_1$ falls on P. Turn the aperture of $S_2$ so that light from $B_2$ does *not* fall on P and is not reflected towards P. Then adjust the distance of $B_1$ from P so that the meter deflection has some convenient value, $\theta_1$, about 10% of full-scale deflection, and record $\theta_1$. *Do not move $B_1$ again* in this part of the experiment. Now arrange the mask M to cut off the light from $B_1$ (see Fig. 41*a*), and adjust the position of $B_2$ until light from it on P gives the *same* deflection $\theta_1$. Remove the mask M (without disturbing either bulb or their shields) and record the new deflection $\theta_2$ due to the light from both bulbs.

Mask $B_1$ again and move $B_2$ until it alone produces a deflection of $\theta_2$. Then unmask $B_1$ and record the deflection $\theta_3$ due to $B_1$ in its fixed position and $B_2$ in its new position. Repeat, moving $B_2$ each time until it alone gives the same deflection as both bulbs at the previous measurement, and then unmask $B_1$. Proceed in this way until the full-scale deflection is reached.

## MEASUREMENTS

| Measurement no. (units of flux) | 1 | 2 | 3 | 4 | 5 | 6 |
|---|---|---|---|---|---|---|
| Deflection $\theta$ from 'zero' | | | | | | |

GRAPH

If the light reaching P from $B_1$ is taken as one unit of luminous flux, giving a deflection $\theta_1$, then $\theta_2$ is the deflection corresponding to two such units, assuming two quantities of flux produce a resultant equal to their sum. $B_2$ was then adjusted to give two units of flux at P, and $B_1$ then added to find the deflection corresponding to three units—and so on.

Thus, if the combination of photocell and meter gives a deflection directly proportional to the incident flux on the cell, a graph of $\theta$ against the number of units of flux will be a straight line through the origin.

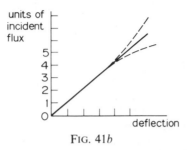

FIG. 41*b*

Plot the deflections $\theta_1$, $\theta_2$, etc., against the integers 1, 2, etc., representing units of flux, as in Fig. 41*b*. Test whether the plotted points are consistent with a straight line through the origin, or whether there is evidence of non-linearity as indicated in the figure by dotted lines.

## 2. INVERSE SQUARE LAW

METHOD

With the photocell still connected as above and using the same value of $R$, set up one (unshielded) bulb B in front of the cell as in Fig. 41*c*. Check that the meter 'zero' reading with the bulb off is unchanged. Now switch on the bulb, and find roughly the range of distance over which it may be moved to give deflections between full scale and about 10% of the full scale.

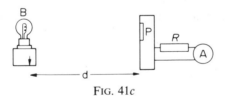

FIG. 41*c*

Covering this range in about 10 equal steps, record pairs of values of the distance $d$ cm from B or a mark on its holder, to the face of the cell and the corresponding deflections $\theta$ of the meter. As before, record the deflections from the 'zero' reading with the bulb switched off.

Stray reflections from the bulb should be avoided in both parts of the experiment; in particular, the experimenter should stand *behind* the photocell when taking readings, and nearby reflecting objects should be removed or shielded with black cloth.

(*Experiment continued overleaf*)

# 41. (*Continued*)

MEASUREMENTS

| Distance $d$/cm | | | | |
|---|---|---|---|---|
| Deflection $\theta$ | | | | |
| $1/\sqrt{\theta}$ | | | | |

GRAPHS

The inverse square law may be tested over the range for which the cell and meter are linear, by plotting a suitable graph. If $\theta \propto 1/d^2$, then $1/\sqrt{\theta} \propto d$. Plot values of $1/\sqrt{\theta}$ v. $d$, as in Fig. 41$d$. Since the position of the active surface of the cell is usually difficult to determine, and readings may have been taken from a point on the bulb holder not in line with the filament, there is a zero error in the values of $d$. This will merely produce a non-zero intercept equal to the zero error.

Test whether the points plotted are consistent with a straight line over the range for which the cell and meter are linear. Discuss whether any discrepancy observed could in fact be due to any non-linearity detected above, or has some other source.

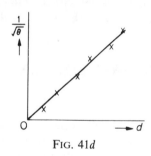

FIG. 41$d$

CONCLUSIONS

With .. $\Omega$ in series, the photocell and meter were linear in their response over a range of deflections from .. to ..

The plotted values of $d$ v. $1/\sqrt{\theta}$ were consistent with the relation .. over the range ..

ERRORS

The procedure described is intended to minimise some sources of error. Discuss these, and justify any alterations or improvements in procedure that you make. Both bulbs are left on throughout (1) to avoid the effect on the brightness of one when switching off the other, if their power supplies are not independent.

In the first half of the experiment, errors occur in setting $B_2$ so that the new illumination is equal to that due to both bulbs previously. Discuss whether such errors are cumulative.

# PART FIVE

# Waves

## EXPERIMENT 42

# Measurement of Wavelength of Red and Blue Light by Young's Fringes

FIG. 42

APPARATUS

12 V motor headlamp bulb M with straight filament, short-focus condensing lens C, adjustable slit $S_1$, double slit $S_2$ (see 'Note' at end), transparent millimetre scale R, eyepiece E, red and blue filters F.

METHOD

At one end of a long bench set up the headlamp bulb with its filament vertical. With the lens C, focus the image of the filament onto the adjustable slit $S_1$, which may be the collimator slit from a spectrometer. Clamp $S_1$ so that the slit is vertical, and place the double slit $S_2$ about 0·5 m from $S_1$. Pick up the light transmitted by $S_2$ on a piece of card, and move the card some 2 m beyond $S_2$. Clamp the eyepiece E (which may be taken from a spectrometer telescope) so that the light seen on the card falls centrally on it. Coloured fringes should now be seen through E. Vary the width of the slit $S_1$ until the fringes are distinct, but not too faint to be seen. Fix the transparent ruler R in front of E so that the millimetre markings are clearly in focus. Place the red filter F in the path of the light, and measure the distance occupied by, say, 10 dark fringes on the ruler R. Repeat using the blue filter.

Measure the distance $y$ cm from the double slit $S_2$ to the plane of the ruler R. Measure the distance $t$ between the centres of the double slits with a travelling microscope.

MEASUREMENTS

Distance occupied by .. fringes = .. mm (red)= .. mm (blue)
∴ distance $x$ between dark fringes= .. mm (red)= .. mm (blue)
Distance $y$ from $S_2$ to R = .. m
Spacing $t$ of double slit = .. mm

CALCULATION

The wavelength $\lambda$, for red and blue light, is given by

$$\lambda = \frac{xt}{y}$$

where $y$ is the distance from the double slits to the ruler, $t$ is the double slit spacing, $x$ is the distance between two dark fringes.

## CONCLUSION

The wavelength for red light was found to be .. m or .. nm
„   „   „ blue „   „   „   „ „ .. m or .. nm

## NOTE

1. $1 \text{ nm} = 1$ nanometre $= 10^{-9}$ m. Thus $5 \cdot 89 \times 10^{-7} \text{ m} = 589 \text{ nm}$.

2. *Construction of double slits.* The slits may be ruled on a slip of glass coated with soot, or (better) 'Aquadag', or on a fogged photographic plate. The slits should be ruled parallel to each other about $0 \cdot 5$ mm apart, and 10 mm long, using a steel ruler and a fine needle or sharp knife. The slits should be of approximately the same width.

3. *Spectrometer use.* Young's fringes can be quickly set up using a spectrometer adjusted for parallel light (p. 82). Narrow the collimator slit, place the double slits on the table so that the light is incident roughly normally to their plane, and observe the slits through the telescope. Fringes should be seen on moving the table slightly for adjustment of the incident light.

Measure the angular separation $\theta$ of say 10 bright bands with the telescope. From diffraction theory, $\sin \theta = 10 \lambda / t$. From your measurement of $\theta$, (a) can the approximation $\theta = 10\lambda/t$ be made? (b) if so, calculate an approximate value for $\lambda$ from $\theta = 10\lambda/t$.

## ERRORS

The largest errors occur in the measurement of the fringe and slit spacing. Measure the error in setting the crosswires by taking several readings. The distance $y$ may readily be measured to, say, $0 \cdot 1\%$.

## ORDER OF ACCURACY

$$\frac{\delta\lambda}{\lambda} \times 100\% = \left[ \frac{\delta t}{t} + \frac{\delta x}{x} \right] \times 100\% \text{ (approx.)}$$

# EXPERIMENT 43

## Measurement of Wavelength of Sodium Light using Newton's Rings

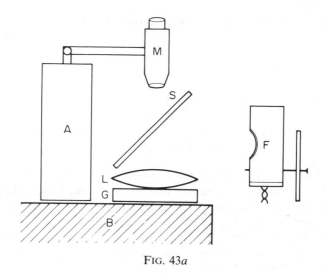

FIG. 43a

APPARATUS

Long-focus lens L, glass block G, thin glass sheet S, travelling microscope M, sodium lamp F, spherometer.

METHOD

Thoroughly clean the lens L and glass block G, both of which should be free from scratches, and of good optical quality. Place L on top of G near the edge of the bench B, and fix a sodium lamp near L, as shown in Fig. 43a. Arrange the base A of the travelling microscope M so that the microscope is vertically above the centre of the lens, and focus the microscope onto the upper surface of G. (A slip of paper inserted between L and G will assist in this.) Fix the sheet of glass S in a clamp (not shown) at about 45°, and insert the sheet between L and M. Looking into the microscope, vary the position and angle of S until the field of view is as bright as possible. If rings are not now visible, slightly raise or lower M until they appear, and focus as sharply as possible on them. Move the microscope sideways until the rings are central in the field of view.

Set the crosswires on the centre of the rings, and use the vernier screw to move the crosswires out to, say, the 10th dark ring. Read the vernier with the crosswires set on the 10th ring, and then bring the microscope back, stopping on each ring and reading the vernier. Continue past the centre, and take readings on the opposite side until the 10th ring is again reached.

Measure the radius of curvature of the *lower* face of the lens by Boys' method (p. 92).

106

MEASUREMENTS

| Radius of curvature of lens $R= ..$ mm | No. $n$ of dark ring | | Micrometer reading/mm | | Diameter $d$/mm | $d^2$/mm$^2$ |
|---|---|---|---|---|---|---|
| | (1) | (2) | (1) | (2) | | |
| | 10 | 10 | | | | |
| | 9 | 9 | | | | |
| | 8 | 8 | | | | |
| | 7 | 7 | | | | |
| | 6 | 6 | | | | |
| | 5 | 5 | | | | |
| | 4 | 4 | | | | |
| | 3 | 3 | | | | |
| | 2 | 2 | | | | |
| | 1 | 1 | | | | |

## CALCULATION

Subtract the micrometer readings corresponding to each ring to obtain the diameters $d$ of the rings, and enter these in the table of measurements, together with $d^2$.

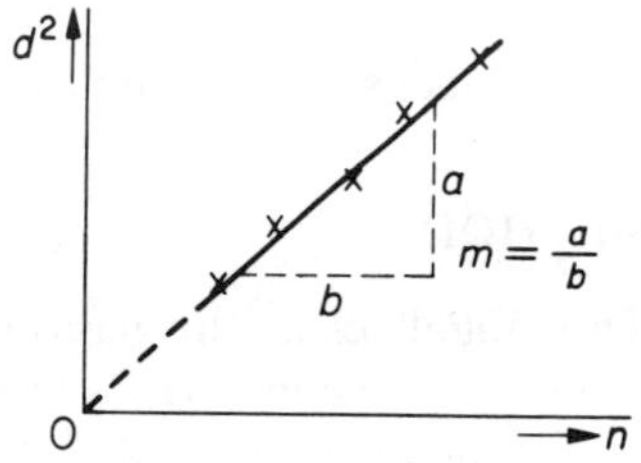

## GRAPH

Plot a graph of $d^2$ v. $n$ (Fig. 43b). For a dark ring, if $r$ is the radius of the ring,

$$r^2 = Rn\lambda$$

FIG. 43b

where $R$ is the radius of the lens face and $\lambda$ is the wavelength of sodium light,

$$\therefore\ d^2 = 4R\lambda n$$

and so the graph is a straight line, passing through the origin, of slope $4R\lambda$.

Calculate from the slope the value of $\lambda$, using the measured value of $R$.

## CONCLUSION

The wavelength of sodium light was found to be .. m or .. nm (see p. 105).

## ERRORS

1. There is error in setting the crosswires on the rings, and in reading the vernier.
2. The screw may suffer from 'backlash' but the method described will tend to eliminate this error.
3. Errors occur in the measurement of $R$.

## ORDER OF ACCURACY

Find the variation possible in the slope by drawing lines of greater and lesser slope that just intersect most points. If the slope is $m$, and the estimated error in it is $\delta m$,

then
$$\frac{\delta \lambda}{\lambda} \times 100\% = \left[\frac{\delta m}{m} + \frac{\delta R}{R}\right] \times 100\%$$

# EXPERIMENT 44

## Measurement of Thickness of Foil and Angle of Air Wedge by Interference Fringes

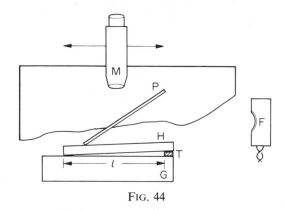

FIG. 44

APPARATUS

Glass blocks G, H, tinfoil slip T, glass plate P, sodium lamp F, travelling microscope M.

METHOD

Thoroughly clean the glass blocks G, H, and place them together with a small slip of tinfoil or paper, T, under one end. Arrange the sodium lamp F beside the block, and focus the travelling microscope on the upper surface of G. Introduce the glass plate P, and alter its position and angle so that the field of view is as bright as possible. Adjust the focus of the microscope so that fringes are visible; they should be straight lines perpendicular to the length of the air film. Turn the microscope stand until the direction of traverse of the microscope is along the air film, and set the crosswires on a dark fringe near one end; read the vernier. Then traverse the microscope along the air film, counting the fringes, and after a number of fringes, N, read the vernier again. Now measure the length $l$ of the air wedge, from the end of H in contact with G to the inner edge of the foil, by means of the microscope.

MEASUREMENTS

$$1\text{st vernier reading} = \text{.. mm}$$
$$2\text{nd} \quad \text{,,} \quad \text{,,} \quad = \text{.. mm}$$
$$\text{Distance } d \quad = \text{.. mm}$$
$$\text{No. of fringes } N = \text{..}$$
$$\text{Length of air film } l = \text{.. mm}$$

CALCULATION

Over a distance $d$, the change in thickness of the wedge is $d \tan \alpha$, where $\alpha$ is the angle of the wedge. Calculate the value of $\tan \alpha$ from the equation

$$N\lambda = d \tan \alpha$$

where $N$ is the number of dark fringes, and $\lambda$ is the wavelength.

Calculate the thickness of the foil $t$ from the equation:

$$t = l \tan \alpha = \ldots \text{ mm}$$

## CONCLUSION

The thickness of the foil $t = \ldots$ m or $\ldots$ mm
Angle of air wedge, $t/l \quad = \ldots$ radian

## ERRORS

(1) Errors in reading the vernier, and hence in the lengths of $d$ and $l$.
(2) Error in setting the crosswires centrally on a dark fringe.

## ORDER AND ACCURACY

$$\text{Maximum percentage error in } t = \frac{dt}{t} \times 100\% = \left[ \frac{\delta d}{d} + \frac{\delta l}{l} \right] \times 100\%$$

Attempt to measure $t$ with the best micrometer available, and account for any discrepancy.

# EXPERIMENT 45

# Measurement of Wavelength by Diffraction Grating

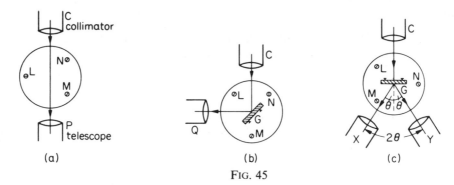

FIG. 45

## APPARATUS

Replica diffraction grating G, spectrometer, sodium discharge tube.

## METHOD

1. *Adjustment.* Focus the telescope and collimator for parallel light. See Expt. 31, p. 82. Turn the telescope to P to view directly the image of the illuminated slit (Fig. 45a). Move the sodium source if necessary until the image is clear and bright; the source need not be very close to the slit. Make the slit narrow and centre the image of the slit exactly on the crosswires. This may be done more accurately if the crosswires are set at 45° to the vertical. Read the vernier, and with the table locked, turn the telescope to Q (Fig. 45b), so that the vernier reading is altered by exactly 90°.

Place the diffraction grating G in its holder on the table, with its plane perpendicular to the line joining two of the screws, L, M (Fig. 45b). Lines are ruled on the table as a guide. (Do *not* touch or attempt to clean the surface of G.) Lock the telescope, and turn the *table* until the image of the slit by reflection at the surface of G appears. Adjust one or both of the screws L, M until the image is in the centre of the field of view. Move the *table* with the slow motion screw until the image is centred on the crosswires. Again read the vernier, and with the telescope locked, turn the table so that the reading changes by exactly 45°, and G is perpendicular to light from C as in Fig. 45c. Lock the table.

Turn the telescope to a position such as X (Fig. 45c), to receive the first-order diffracted image. If the grating lines are not perpendicular to the plane of rotation of the telescope the image may be displaced from the centre of the field of view. Adjust the third screw N until the image is central.

2. *Angle of diffraction.* With the telescope at X, and the slit as narrow as possible, set the crosswires in turn on the two closely spaced components $D_1$, $D_2$, of the sodium $D$ line. Record the vernier readings corresponding to the $D_1$ and $D_2$ lines. The $D_1$ line is (conventionally) that of longer wavelength: that is, the more deviated line. Turn the telescope to Y, on the other side of the incident light, and repeat the measurements.

Look for the second-order diffraction images (at larger angles), and record vernier readings for settings on the $D$ lines on either side of the normal for this order.

## MEASUREMENTS

1. *Adjustments.* Reading with telescope at P = . .     Alter by 90° = . .
Reading with telescope at Q
and table adjusted          = . .     Alter by 45° = . .

110

## 2. Diffraction angle.

| | Vernier angles | | $2\theta$ | $\theta$ | Vernier angles | | $2\theta$ | $\theta$ |
|---|---|---|---|---|---|---|---|---|
| | At X | At Y | | | At X | At Y | | |
| Line $D_1$ | | | | | | | | |
| Line $D_2$ | | | | | | | | |

Grating spacing: Lines per cm= ..

### CALCULATION

Find the differences $2\theta$ between corresponding pairs of vernier readings on either side of the normal (at X and at Y) for the $D_1$ and $D_2$ lines in the first- and second-order diffracted images. Enter these and the angles of diffraction in the tables as indicated.

Calculate the grating spacing $d=1/$no. of lines per cm= .. cm= .. m.

Calculate the wavelengths of the $D_1$ and $D_2$ lines, substituting appropriate values in:

| | First Order | Second Order | Average |
|---|---|---|---|
| $D_1$ | $\lambda_1 = d\sin\theta_1$ <br> $\lambda_1 = $ .. m | $2\lambda_1 = d\sin\theta$ <br> $\lambda_1 = $ .. m | $\lambda_1 = $ .. m |
| $D_2$ | $\lambda_2 = d\sin\theta$ <br> $\lambda_2 = $ .. m | $2\lambda_2 = d\sin\theta'$ <br> $\lambda_2 = $ .. m | $\lambda_2 = $ .. m |

### CONCLUSION

The wavelengths of the two sodium lines were found to be:

$$\lambda_1 \,(D_1)= \text{ .. } \times 10^{-7}\text{m or .. nm (see p. 105)}$$
$$\lambda_2 \,(D_2)= \text{ .. } \times 10^{-7}\text{m or .. nm}$$

### ERRORS

1. There is error in setting the crosswires on the image, which may be estimated by measuring the displacement when a setting is made just off centre.

2. Error arises in reading the vernier: note the accuracy with which this can be done.

3. What error is minimised by taking readings on either side of the normal?

4. Replica gratings are made by stripping a collodion or plastic film off the surface of a ruled grating, whose spacing is that recorded. Systematic error may thus arise from distortion of the film, during stripping or when it is mounted on glass.

### ORDER OF ACCURACY

Find from the tables the errors $\delta(\sin\theta)$ in values of $\sin\theta$ caused by the total error $\delta\theta$ in 1. and 2. above. The order of accuracy is given by:

$$\frac{\delta\lambda}{\lambda} \times 100\% = \frac{\delta(\sin\theta)}{\sin\theta} \times 100\%$$

Discuss whether any unexplained discrepancy with the accepted values may be due to systematic error as in 4. above.

### PROBLEM

Use a section cut from a 45 r.p.m. gramophone record as a *reflection* grating and measure the wavelength of light passed by red and blue filters.

# EXPERIMENT 46

## Resolving Power of Telescope

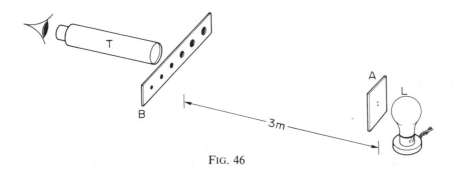

FIG. 46

### APPARATUS

Telescope T (with achromatic objective, e.g. spectrometer telescope); 100 W opal (mains) lamp L (or 12 V–24 W lamp and diffusing screen); cardboard photographic mounts, aluminium (cooking) foil, needle; opaque strip B, aluminium or paxolin, drilled with holes varying from 3 mm to 1 mm diameter about 30 mm apart; travelling microscope.

### METHOD

1. Place a piece of foil A in a cardboard photographic mount and make a small hole in it with a needle. By means of the lamp L, illuminate the hole from behind (Fig. 46). With the telescope 3 m or more from A, focus as sharply as possible on the hole in the foil. (If a narrow filament lamp is used for illumination, the telescope will have to be moved until the light passing through the hole falls on it.) If the light is very intense, it may be difficult to focus sharply (*why?*). Try reducing the intensity, perhaps with a diffusing screen, until the hole can be seen clearly.

2. Now place the drilled strip B in front of the telescope objective, so as to reduce the aperture of the lens (see Fig. 46). Start with the largest hole, and then proceed to smaller and smaller apertures, when both the brightness and form of the image are both altered. Describe the change in brightness.

The image seen depends on the shape of the object hole (is it circular?) and on the shape of the telescope aperture hole (is that circular?). If both are circular, theory suggests that a diffraction pattern, composed of a central bright region surrounded by bright and dark rings, will appear. Look carefully at the size of the image, and at the region round it, as the aperture of the telescope is reduced when different holes in B are used. Describe what you see, paying attention to the *size* of the central bright region which appears.

3. Now make a second hole, about 1 mm from the first, in the foil A. (Using a vertical filament lamp, it is best to have the holes in a vertical line so that both are illuminated.) Check that both holes can be seen as two clear 'stars' in the telescope, and adjust the positions of the lamps and the holes until both are equally illuminated.

If the drilled strip B is now used to observe them, the holes will probably be distinguishable with the largest aperture. Through the smallest aperture, they may appear to be one patch of light surrounded by rings. If they can be distinguished in the latter case, however, try holes closer together until one patch of light is seen using the smallest aperture.

Now find the particular aperture for which the two holes in A can just, but only just, be distinguished. The holes are then said to be just *resolved*. Measure (i) the diameter of this aperture, (ii) the spacing of the holes in A, (iii) the distance from the holes to the telescope objective.

## MEASUREMENTS

Spacing of illuminated holes (centre to centre), $h=$ .. mm
Distance of holes to telescope objective, $y$     = .. mm
Least aperture diameter for resolving holes, $D$    = .. mm

## COMPARISON WITH THEORY

Diffraction theory suggests that the two holes are just resolved when the central bright region of one ring pattern just falls on the dark outer edge of the central bright region of the other. Theory then shows that the angle $\delta\theta$ subtended at the objective by the two holes has now an order of magnitude $\lambda/D$, where $\lambda$ is the wavelength of the light and $D$ is the telescope aperture. Take the mean value of $\lambda$ as $6 \times 10^{-4}$ mm and calculate $\delta\theta$.

$$\delta\theta \text{ (theory)} \approx \frac{\lambda}{D} = \text{ .. radian}$$

Now calculate $\delta\theta$ from the table of measurements:

$$\delta\theta \text{ (measurements)} = \frac{h}{y} = \text{ .. radian}$$

Compare the two values for $\delta\theta$.

## PROBLEM

Place red, and then blue, colour filters behind the holes in A. See whether the holes can be resolved more easily with one colour of light than the other.

The wavelength of blue light is perhaps 1·5 times less than that of red, depending on the filters. What would you predict from theory? Would you expect any difference to be observable?

## ADDITIONAL

*Resolving power of eye.* View a well-illuminated rule which is graduated in millimetres. Move back from the rule until two adjacent millimetre graduations are *just* not resolved. Then measure your distance $x$ to the rule. The approximate resolving power of the eye (the smallest angle which can just be resolved) is given by $1\,\text{mm}/x\,\text{mm}$.

Assuming the mean wavelength $\lambda$ of light is $6 \times 10^{-4}$ mm and that the resolving power is given approximately by $\lambda/d$, where $d$ is the diameter of the eye-pupil, estimate a value for $d$ under the particular conditions of illumination.

EXPERIMENT 47

# Interference and Other Experiments with Microwaves

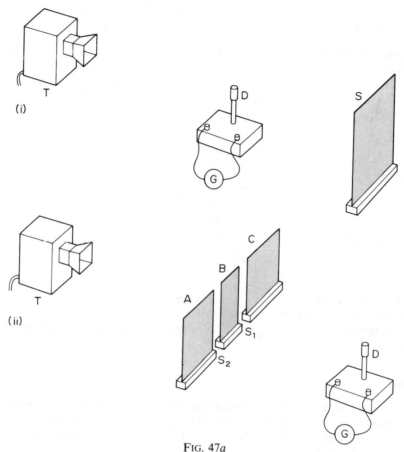

FIG. 47a

## APPARATUS

Microwave transmitter T and power supply; probe with microwave diode D; galvanometer G (*Scalamp* or $0-100\,\mu A$) and/or audio-amplifier and loudspeaker; metal sheet S about 30 cm square (Fig. 47a (i)); metal sheets A, B, C to form slits $S_1$, $S_2$ about 2 cm wide, 6 cm apart (Fig. 47a (ii)).

## METHOD

### 1. Use of Apparatus

Connect the transmitter T to the power supply, following the manufacturer's instructions (Fig. 47a (i)). Place a diode 'probe' D, which may be mounted on a box with terminals, in front of T and connect it to the sensitive meter G. From observations, what evidence have you that the transmitter is sending some energy to D across the space between them? Find out whether a book, wood, metal, glass, water and other materials transmit the radiation from T.

Devise an experiment to see if any materials *reflect* the radiation. If so, does the angle of reflection equal the angle of incidence? Does your hand transmit or reflect the radiation?

Instead of a meter G, the diode D can be connected to an audio-amplifier with a loudspeaker. The microwave oscillations are of extremely high frequencies and hence cannot be heard. Most transmitters, however, can be 'modulated' to provide an audio-frequency signal, usually by being switched on and off 100 or 1000 times per second, and this can be heard. The loudness of the sound indicates the intensity of the transmitted signal. With the diode D, explore the variation in direction of the strength of the beam from the transmitter, going as far away from it as possible. The signal may rise and fall as D is moved away, especially if the beam is pointing at a wall or other obstruction. Can you explain this? The experiment now to be described makes this effect happen deliberately.

## 2. Stationary (Standing) Waves

Place a metal sheet S about 50 cm from the transmitter T. Does S reflect the transmitter beam? Turn S to reflect the beam back to T (Fig. 47a(i)). Now move the diode D in the space between S and T, both sideways and lengthways.

Mark the positions of the diode D whenever the signal is very faint. Are such positions regularly spaced or not? Record the average distance between two such successive positions, measuring the length occupied by 10 or more.

Now leave D in one place and move the metal sheet S towards and away from D. Again find the different positions of S when the signal received by D is very faint. How are these positions spaced? Compare your result with the previous value, when D was moved and S was fixed.

MEASUREMENTS

Average spacing of weak signal positions—

$$\text{diode moved} \quad = \text{.. mm}$$
$$\text{metal sheet moved} = \text{.. mm}$$

CALCULATION

If the radiation is a wave and reflection occurs at the sheet S, then a stationary (standing) wave is set up between the transmitter and S. This is due to two waves travelling in opposite directions. The distance apart of the *nodes*, where the signal is very faint, is $\lambda/2$, where $\lambda$ is the wavelength. Calculate $\lambda$.

$$\lambda = \text{.. mm}$$

## 3. Interference of Waves

Arrange the transmitter T about 50 cm behind the two slits $S_1$, $S_2$, about 2 cm wide and 6 cm apart, using the metal sheets A, B, C as shown in Fig. 47a(ii). Place the diode D by eye on the centre line from T through the middle of B and about 50 cm from $S_1$, $S_2$. The signal should now be strong; adjust the position of D slightly until it is at its strongest. The audio-amplifier is most sensitive for this purpose.

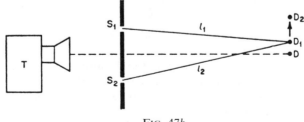

FIG. 47b

*(Experiment continued overleaf)*

115

# 47. (*Continued*)

Now slowly move the diode D sideways, parallel to the plates, and find the first position $D_1$ where the signal is weakest (Fig. 47b). Measure the respective distances $l_1$, $l_2$ from $D_1$ to the middle of $S_1$, $S_2$. Move the diode sideways to the next position such as $D_2$ where the signal is again faintest, and measure the new lengths $l_1$, $l_2$.

Try moving the diode *towards* the slits $S_1$, $S_2$, keeping the signal faint all the time, as it was at $D_2$. Measure the new value of $l_1$ and $l_2$ for a position closer to the slits.

MEASUREMENTS

| $l_1$ /mm | $l_2$ /mm | $(l_2 - l_1)$ /mm |
|---|---|---|
|  |  |  |

CALCULATION

Calculate the difference $(l_2 - l_1)$ for each faint signal position of the diode D. Here the waves from one slit must be arriving at D exactly opposite in phase to that arriving from the second slit. The resultant signal, obtained by adding the two waves, is then small. *If* the waves from T reach the slits $S_1$, $S_2$ in phase, and $\lambda$ is the wavelength, then $(l_2 - l_1)$ should be $\lambda/2$, or $3\lambda/2$, or any odd number of half wavelengths. Check whether all your results for $(l_2 - l_1)$ fit this conclusion. (Compare *Young's experiment for light waves*, p. 104.) Why is it difficult to do this experiment accurately?

## FURTHER EXPERIMENTS

1. *Diffraction.* Place a metal sheet S so that it *nearly* cuts off the signal from the transmitter T to the diode D, as in Fig. 47c. Now place a second sheet S' near S, so as to remove *more* of the waves going round the edge of S to D. When a narrow slit is left between S and S' the signal should increase! Can you explain this? What property of microwaves does this experiment show?

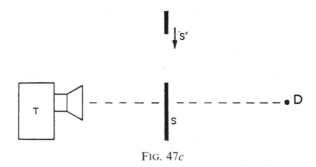

FIG. 47c

2. *Lloyd's mirror—Interference.* Place the transmitter T close to the plane of a vertical metal sheet S (Fig. 47d). Move the probe D near S, and adjust T so that the probe receives waves directly from T and also by reflection at S, as shown. Alter the positions of T and D to obtain the minimum signal at D. In this case destructive interference occurs at D. Now alter the position of D slightly to obtain a maximum signal or constructive interference.

The waves reflected from S appear to come from the 'mirror image' of T in S. By measuring the distance from this mirror image to D, and the distance TD, estimate roughly the wavelength.

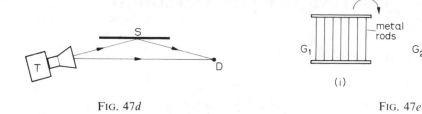

FIG. 47d                    FIG. 47e

3. *Polarization.* Place the probe directly in front of the transmitter so that a large signal is received by the probe, as shown by the deflection in the galvanometer.

Now introduce a metal grille $G_1$ between the transmitter and the probe so that the rods in the grille are vertical (Fig. 47e(i)). Observe any change in the strength of the signal now received by the probe. Then rotate the grille *in its own plane* until the rods are horizontal, as $G_2$ in Fig. 47e(ii), and at the same time observe the signal received by the probe.

(*a*) Why can you say that the waves from the transmitter are 'plane-polarized'? (*b*) In which plane are the waves plane-polarized, that is, which plane contains the electric vibrations? Confirm your answer to (*b*) by rotating the transmitter in its own plane with the grille positioned as $G_1$ in Fig. 47e(i) and observing the strength of the signal received by the probe.

# MECHANICAL WAVES AND SOUND

## EXPERIMENT 48

# Speed of Waves on Springs

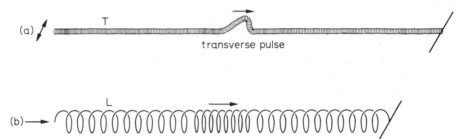

(a) transverse pulse

(b) longitudinal pulse

FIG. 48

APPARATUS

Long narrow spring, about 3 m long, 20 mm diameter; long 'Slinky' spring of narrow flat steel strip, in a coil about 100 mm diameter; stop-watch; metre rule; top-pan balance to weigh the springs; spring balance 0–50 N.

METHOD

### A. *Transverse pulse* (*wave*) *in long narrow spring*

1. Tie one end of the long spring T firmly to something like a table leg and lay out the spring on a *smooth* floor—if necessary, find a suitable place to do the experiment.

2. Stretch the spring, and practice sending transverse pulses along the spring, as shown in Fig. 48a, by giving one end a sharp, sideways shake. Notice what happens to the pulse when it is reflected from the fixed end. It should be possible, if the free end is held firmly in the hand, to get the pulse to travel four or five times along the spring.

3. Now time the pulse as best as you can, counting the number of times it travels along the spring in that time. Record the measurements and repeat them.

Now measure the *stretched* length of the spring and the tension in newtons in the spring using a spring balance. Record the results.

4. Next, oscillate the end of the spring continuously, at such a rate that a *standing wave* with two loops (a node at each end and one in the middle) is produced. Time about 20 of these oscillations and record it.

5. Now stretch the spring until the pulse visibly travels faster, and repeat all the above measurements.

6. Finally, weigh the spring (coiled up) and record its mass.

118

## MEASUREMENTS

| | |
|---|---|
| Time of travel of pulse along spring, $t$ | .. s, .. s |
| No. of times pulse travels along spring, $n$ | .. , .. |
| Length of stretched spring, $L$ | .. m |
| Tension in spring, $T$ | .. N |
| Mass of spring, $M$ | .. kg |
| Time for .. oscillations for 1 wavelength standing wave, $t_s$ | .. s |

## CALCULATIONS

(i) Calculate the measured speed of a transverse pulse at each tension—use the fact that the pulse travels a distance $nL$ in time $t$, so that

$$v = \frac{nL}{t} = \text{ .. m s}^{-1} \text{ (direct method)}$$

(ii) Obtain a second value for $v$ from the standing wave data. Here the wavelength $\lambda$ (two half wavelengths) is $L$, and the frequency $f$ can be found from the time for several oscillations. Then

$$v = f \times \lambda = \text{ .. m s}^{-1} \text{ (standing wave)}$$

(iii) Now calculate the theoretical speed. For a transverse wave,

$$v = \sqrt{\frac{T}{\mu}} = \text{ .. m s}^{-1}$$

where $T$ is the tension in newton and $\mu$ is the mass in kg per metre length of the stretched spring, that is, $\mu = M/L$.

(*Experiment continued overleaf*)

## 48. (*Continued*)

### B. *Longitudinal pulse (wave) in 'Slinky' spring*

If you wish, you can repeat the previous observations of transverse waves using a 'Slinky' spring. However, the 'Slinky' will also carry longitudinal waves.

1. Practice sending a longitudinal compression pulse along the stretched 'Slinky' L, as in Fig. 48*b*, by giving one end a sharp push forward. Notice what happens to the pulse when it is reflected from the far end, which should be firmly fixed as before. Try also rarefaction pulses (pull the end back sharply).

2. Time the pulses along the spring, noting the number of times the pulse travels the length of the spring. Record the results.

Measure the length of the stretched spring. Now use a spring balance to measure the *increase* in tension when the spring is stretched by a length $x$ from an initial length $y$.

Finally, weigh the spring and record its mass.

3. Stretch the spring more and repeat all the above measurements.

MEASUREMENTS

| | | |
|---|---|---|
| Time of travel of pulse along spring, $t$ | . . s, . . s | |
| No. of times pulse travels along spring, $n$ | . . , . . | |
| Length of stretched spring, $L$ | . . m | |
| Mass of spring, $M$ | . . kg | |

Force $F$ increases by . . N for extension $x$ of . . m from previous length $y$ of . . m

## CALCULATIONS

(i) Calculate the measured speed of a longitudinal pulse on the 'Slinky' as before, using time and distance.

$$\text{Speed } v = \frac{\text{distance}}{\text{time}} = \text{ .. } \mathrm{m\,s^{-1}} \text{ (direct method)}$$

(ii) Calculate a theoretical value of the speed. The speed of a longitudinal wave on such a spring is given by

$$v = \sqrt{\frac{k}{\mu}}$$

where $\mu$ is the mass per metre of the stretched spring and $k$ is a 'spring constant' given to a good approximation by $k = Fy/x$, where $F$ is the increase in force needed to stretch the initial length $y$ of the spring by a length $x$, measured above.

From the measurements, obtain $k$. With the value of $\mu$ in $\mathrm{kg\,m^{-1}}$, calculate $v$.

$$v = \text{ .. } \mathrm{m\,s^{-1}} \text{ (theoretical value)}$$

## CONCLUSIONS

The speed of the transverse pulse was .. $\mathrm{m\,s^{-1}}$.
The speed of the longitudinal pulse was .. $\mathrm{m\,s^{-1}}$.

## DISCUSSION

Consider whether there is evidence that the theories correctly predict the wave speeds. Point out any features of the experiments which may give rise to disagreement with the theoretical values.

# EXPERIMENT 49

## Investigation of Variation of Frequency of Stretched String with Length

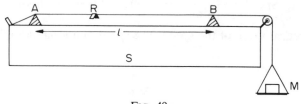

FIG. 49a

### APPARATUS

Sonometer S with wire AB passing over grooved pulley, bridges A, B, six tuning forks of known frequencies, masses M of several kilograms.

### METHOD

Attach a mass M of several kilograms to the free end of the wire; keep M constant, so that the wire has a constant tension. Place the bridges A, B under the wire, and keeping A fixed, adjust the position of B so that the wire AB is tuned to the tuning fork of lowest frequency. M should be such that AB is not less than two-thirds of the available length of wire. At resonance: (i) a small paper rider R in the middle of AB is agitated violently when the tuning fork is sounded and its shank is placed on B or on the sonometer box; (ii) very slow 'beats' are heard between the notes from the plucked wire AB and the sounding tuning fork. Method (i) or (ii) should be used for the final adjustment after tuning the wire roughly by ear. In this way, obtain the six lengths $l$ of the wire which have the same frequencies as the respective forks.

### MEASUREMENTS

Tension (constant)= .. N

| $f$/Hz | | | | | | |
|---|---|---|---|---|---|---|
| $l$/cm | | | | | | |
| $\dfrac{1}{l}$/cm$^{-1}$ | | | | | | |

122

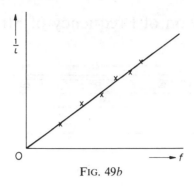

FIG. 49b

Calculate the reciprocals, $1/l$, of the lengths $l$. Plot $1/l$ v. $f$, starting from zero on both axes (Fig. 49b). Test whether a straight line can be drawn from the origin O to lie evenly between the plotted points.

CONCLUSION

The frequency of the note from a plucked string at constant tension varies ...

ERRORS

Errors occur in the measurement of the lengths $l$. Much more important is the uncertainty in setting the bridge in the final tuning. This can be found by measuring the distance the bridge can be moved before the note from the wire goes perceptibly out of tune with the fork, or causes less vibration of the paper rider. The error in $1/l$ can then be found from tables, and plotted on the graph as explained on p. 9.

# EXPERIMENT 50

## Investigation of Variation of Frequency of Stretched String with Tension

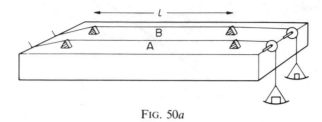

FIG. 50a

### APPARATUS

Sonometer with two wires, A and B, scale-pan and masses, $M$, up to 15 kg.

### METHOD

Place a load of about 2 kg in the scale-pan attached to the wire A, and a fixed load of about 4 kg in the pan attached to B. Adjust the distance apart of the bridges beneath A so that the wires A and B sound approximately the same note when the bridges beneath wire B are well apart, say two-thirds of the total length of B. Do not alter the position of the bridges beneath A during the rest of the experiment, or the weight attached to B. Adjust the vibrating length $l$ of wire B so that it sounds exactly the same note as does A when both are plucked. The most convenient method is to tune B first by ear, and then to listen for beats between the two notes. Vary the length $l$ of B in small steps, and reduce the 'beating' to as slow a rate as possible.

Increase the load on the wire A to about 3 kg, and find the new length $l$ of B which is now in unison with the unchanged length of A. Repeat for other loads up to about 12 kg, recording the load on A and the length $l$ of B. Increase the load by about 50% each time.

In the table of measurements, enter the *tension T* in the wire A in newton (N) for each load of mass $M$—use $T = Mg$ and assume $g = 10\ \mathrm{m\,s^{-2}}$.

### MEASUREMENTS

| Tension $T$/N (fixed length of $A$) | | | | | | |
|---|---|---|---|---|---|---|
| Length $l$/cm ($B$) | | | | | | |
| $\sqrt{T}/\mathrm{N^{1/2}}$ | | | | | | |
| $(1/l)/\mathrm{cm^{-1}}$ | | | | | | |

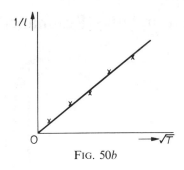

FIG. 50*b*

Calculate the values of $1/l$ and $\sqrt{T}$ for each pair of results, and enter them in the table of measurements.

Plot a graph of $1/l$ v. $\sqrt{T}$, and find whether a straight line may be drawn through the origin among the plotted points (Fig. 50*b*).

If this is the case, then, since $f \propto 1/l$ for a wire under fixed tension (Expt. 49, p. 122), we may deduce that $f \propto \sqrt{T}$ for the wire A.

### CONCLUSION

The frequency $f$ of a fixed length of wire is related to the tension $T$ by an equation of the form …

### ERRORS

1. Error in measuring the length $l$ of wire B.
2. Error in setting the length $l$ of B in unison with A.
3. Error in the tension $T$ due to friction.
4. Variations in the tension of B if it is fixed at both ends, as in some apparatus.

### ORDER OF ACCURACY

Find from tables the errors in $1/l$ and $\sqrt{T}$ caused by the estimated errors $\delta l$ and $\delta T$, and draw short lines through each plotted point parallel to the $1/l$ and $\sqrt{T}$ axes equal in length to the respective errors in $1/l$ and $\sqrt{T}$. If a straight line can be drawn through the origin intersecting each arm of the crosses, the plotted values are consistent with the assumption that $1/l \propto \sqrt{T}$.

# EXPERIMENT 51

# Resonance of Air Column in Tube (Resonance Tube)

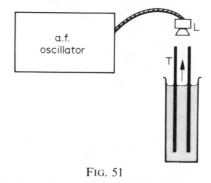

FIG. 51

## APPARATUS

Resonance tube arrangement T, audio-frequency oscillator, suitable earphone or loudspeaker L, thermometer.

## METHOD

1. Adjust the air column in the resonance tube T to its shortest length. Set the a.f. oscillator to a frequency of 300 Hz and adjust the volume control to a low but audible value. Hold the earphone or loudspeaker L a few cm above the open end of the tube T and slowly increase the length of the air column by raising the tube out of the water.

When the first resonant length of air has been obtained, that is, the air in the tube emits the loudest sound, measure the length $l$ of the air column from the top of the tube T. Record the frequency of the oscillator. Determine the uncertainty in the resonant position by adjusting the length of the air column above and below what you consider is the position of maximum sound. Record the diameter $D$ of the tube T.

2. Using different frequencies from 300 to 600 Hz, obtain a series of readings of $l$ which gives resonance of the air column.

3. *Two resonant lengths at the same frequency.* Set the oscillator at a frequency of 320 Hz. As before, determine the resonant length $l_1$ from the top of T. Now increase the length of the air column. At some greater length $l_2$, about three times $l_1$, resonance is again obtained. Record $l_1$ and $l_2$ and their uncertainties.

Repeat with a frequency of 512 Hz in place of 320 Hz.

Record the temperature of the air in the room.

## MEASUREMENTS

*Single resonant length*

| Frequency/Hz | Length of air column /cm |
|:---:|:---:|
| 300 | .. $\pm$ .. |
| .. | .. $\pm$ .. |

Diameter of resonance tube, $D=$ .. cm

126

*Two resonant lengths*

| Frequency /Hz | $l_1$ /cm | $l_2$ /cm |
|---|---|---|
| 320 | .. ± .. | .. ± .. |
| 512 | .. ± .. | .. ± .. |

Air temperature = .. °C

## CALCULATIONS

*Single resonant length*

$$l + c = \frac{\lambda}{4} = \frac{V}{4f}$$

where $c$ is the end correction of the tube T, $V$ is the velocity of sound in the tube and $f$ is the frequency.

*Plot a graph* of $l$ against $1/f$. From the best straight line obtained, deduce (*a*) a value of $V$ from the gradient, which is $V/4$ from above, (*b*) a value for $c$, which is the negative intercept on the axis of $l$ from above.

*Two resonant lengths*

At $l_1$ the length of the resonant column is $\lambda/4$; at $l_2$ for the same frequency, the length is $3\lambda/4$. Eliminating the end correction $c$ by subtraction, we obtain

$$l_2 - l_1 = \frac{\lambda}{2} = \frac{V}{2f}$$

So

$$V = 2f(l_2 - l_1)$$

Calculate $V$ from the readings obtained with the two frequencies.

## CONCLUSIONS

The velocity of sound in air at .. °C is .. m s$^{-1}$.
The end correction of the resonance tube is .. × D.

## ORDER OF ACCURACY

The percentage error in $V$ is the sum of the percentage error in $(l_2 - l_1)$, obtained by adding the actual errors in $l_1$ and $l_2$ and finding the percentage error of $(l_2 - l_1)$, and the percentage error in $f$, which may be obtained by counting the beats obtained with a tuning fork.

Compare your value for $V$ with that given in standard Tables ($V \propto \sqrt{T}$, where $T$ is the absolute temperature). Comment on the accuracy obtained.

Compare your value for the end correction $c$ with the accepted value $c = 0.3$ D approximately, which has been obtained by theoretical arguments.

# EXPERIMENT 52

## Measurement of the Frequency of the A.C. Mains using a Sonometer

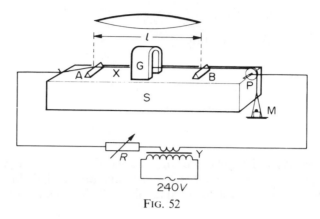

FIG. 52

### APPARATUS

Sonometer S with 24 s.w.g. wire X, 'horseshoe' magnet G, masses $M$ up to 1 kg, rheostat $R$, mains transformer Y (240 V → 6 V), spring balance, chemical balance.

### METHOD

Clean the wire X where it passes over the metal pulley P, and at the far end where it is attached to the sonometer S. Connect the wire, taking one lead from the end and one from the pulley, in series with the rheostat $R$ and the 6 V secondary of the mains transformer Y. Place the magnet G so that the wire passes between its poles, and adjust the current (which may be checked with an ammeter) until the wire can be felt to vibrate slightly, without becoming appreciably heated.

   Weigh the scale-pan on the spring balance. Place a load of about 250 g in the scale-pan, and adjust the positions of the bridges A and B until the wire between them resonates in its fundamental mode, the magnet being at the centre of AB. Measure the length $l$ of the vibrating wire between A and B, and record the combined mass $M$ of the load and scale-pan. Increase the load by about 100 g, and measure the new resonating length. Repeat for a third load. To find the mass per metre, $m$, of the wire, cut a known length $x$ (at least 20 cm) of the wire, and weigh it carefully. If more convenient, $m$ can be found by measuring the diameter of the wire at three different places in perpendicular directions, and obtaining the density of the material from physical tables.

### MEASUREMENTS

| | (1) | (2) | (3) |
|---|---|---|---|
| Total load $M$ /kg | | | |
| Resonating length, $l$/m | | | |

| | | |
|---|---|---|
| Length of wire, $x$/m | | OR   Diameter of wire, $d = \ .. \ $ m |
| Mass of $x$ m of wire /kg | | Density of wire, $\rho \ = \ .. \ \mathrm{kg\,m}^{-3}$ |
| Mass per m of wire, $m$/kg m$^{-1}$ | | Mass per m, $m = \dfrac{\pi d^2 \rho}{4} = \ .. \ \mathrm{kg\,m}^{-1}$ |

128

## CALCULATION

Calculate the mass $m$ in kg of 1 metre of the wire.

Since the wire resonates, the mains frequency $f$ = the frequency of vibration of the string. Calculate the value of $f$ from the values of $M$ and $l$, with $M$ in kilogram, $l$ in metre (m), $m$ in kilogram metre$^{-1}$ (kg m$^{-1}$) and $g = 9 \cdot 8 \, \text{m s}^{-2}$:

$$(1)\ f = \frac{1}{2l}\sqrt{\frac{Mg}{m}} = \ .. \ \text{Hz}$$

$$(2)\ f = \ .. \qquad = \ .. \ \text{Hz}$$

$$(3)\ f = \ .. \qquad = \ .. \ \text{Hz}$$

$$\text{Average } f = \ .. \ \text{Hz}$$

## CONCLUSION

The frequency of the mains was found to be .. Hz.

## ERRORS

1. The tension in the wire may be lower than $Mg$, due to friction at the pulley.
2. Error in detecting the resonant condition. Find how far one bridge has to be moved so as *just* to reduce the amplitude of oscillation.
3. Errors in the length $x$, and the mass of the cut wire.

## ORDER OF ACCURACY

An estimate of the order of accuracy can be obtained by recalculating $f$, using values of $l$, $M$, $g$ and $m$ which combine together to make $f$ as large, or as small, as possible taking account of the errors.

# EXPERIMENT 53

# Measurement of Velocity of Sound in Rod by Dust (Kundt's) Tube

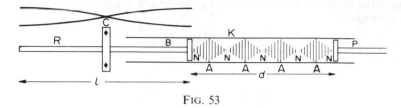

FIG. 53

## APPARATUS

Kundt's tube K, rod R with disc B attached, plunger P, clamp C, metre rule, lycopodium powder or cork dust.

## METHOD

Thoroughly clean and dry the wide glass tube K, and introduce a *thin* layer of lycopodium powder along its length. The powder may conveniently be sprinkled on a metre rule, which is then inserted into the tube and turned over. Clamp the rod R at its mid-point by means of C, with the disc B well inside the tube K, but not touching its walls. The rod is made to sound by drawing a friction pad from the clamp towards the free end. For a glass rod, use a wetted, loose-fitting rubber bung; for steel use a wet cloth; for wood use a dry rosined cloth. As R is made to sound a clear, loud note, move the plunger P slowly into the tube until the motion of the powder is a maximum. The powder will settle with clearly defined nodes N (between which will be striations due to higher harmonics). Note that there are nodes at B and at P. Count the number $n$ of antinodes in the tube between B and P, and measure the distance $d$ between the inner faces of B and P. Measure the length $l$ of the rod, and the temperature of the room.

## MEASUREMENTS

$$
\begin{aligned}
\text{Length of rod } l &= \text{..  m} \\
\text{Length of air column } d &= \text{..  m} \\
\text{No. of antinodes } n &= \text{..} \\
\text{Room temp. } t &= \text{..  °C}
\end{aligned}
$$

## NOTES

If no motion can be observed: (i) Re-dry the tube K and the powder. (ii) Turn the tube so that the powder line is just about to slip downwards before sounding G. (iii) Face P with bakelite (or some smooth surface) to reduce absorption. (iv) Check that B is not touching the walls of the tube.

## CALCULATION

1. Calculate the velocity of sound $V$ in air at room temperature $t\,°C$ from the equation

$$\frac{V}{V_0} = \sqrt{\frac{t+273}{273}}$$

where $V_0$ is the known velocity of sound in air at $0\,°C$ ($331{\cdot}5\,\mathrm{m\,s^{-1}}$).

$$V = \text{..  m\,s}^{-1}$$

2. Calculate the wavelength $\lambda_a$ in air of the note sounded from the equation:

$$\frac{\lambda_a}{2} = \frac{d}{\text{no. of antinodes, } n}$$

$$\lambda_a = \text{ .. m}$$

3. Calculate the frequency $f$ of the note from the equation:

$$f = \frac{V}{\lambda_a} = \text{ .. Hz}$$

4. Calculate the velocity of sound $V_r$ in the rod using the equation:

$$\text{Wavelength in rod} = 2l = \frac{V_r}{f}$$

$$\therefore \ V_r = 2l.f = \text{ .. ms}^{-1}$$

## CONCLUSION

The velocity of sound in the rod was .. $\text{ms}^{-1}$.

## ERRORS

1. Error in setting the plunger P for maximum motion of the powder. Measure how far P has to be moved to cause a just appreciable diminishing of the motion.
2. Error in measuring the distance $d$.
3. Error in measuring the distance $l$.

## ORDER OF ACCURACY

The maximum % error in $V_r$ is given by:

$$\frac{\delta V_r}{V_r} \times 100\% = \left[\frac{\delta d}{d} + \frac{\delta l}{l}\right] \times 100\%$$

## NOTE

From the density $\rho$ of the material of the rod, the Young modulus $E$ for the rod can be found from

$$V_r = \sqrt{\frac{E}{\rho}}$$

$E$ is in $\text{Nm}^{-2}$ when $V_r$ is in $\text{ms}^{-1}$ and $\rho$ in $\text{kgm}^{-3}$.

# EXPERIMENT 54

# Calibration of Audio Oscillator against Mains Frequency

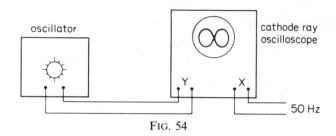

FIG. 54

## APPARATUS

A.F. oscillator, cathode-ray oscilloscope, suitable mains transformer.

## METHOD

If the oscilloscope is not provided with an internal 50 Hz voltage which may be applied to the X plates, connect the X plates to a transformer, which provides a voltage sufficient to produce a horizontal trace about two-thirds of the width of the screen. Connect the oscillator to the Y plates, and switch on. Having allowed some minutes for the oscillator to warm up, when its frequency will no longer drift, set the frequency control near 50 Hz, and adjust carefully until a stationary ellipse is obtained on the oscilloscope screen. Note the reading on the oscillator scale. Increase the frequency to about 100 Hz, and again adjust until a stationary two-looped pattern is formed, and then read the oscillator scale. Repeat for other integral multiples of 50 Hz, up to about 1000 Hz, or as far as is possible with the oscilloscope well focused.

## MEASUREMENTS

| Reading on scale /Hz | | | | | | | | |
|---|---|---|---|---|---|---|---|---|
| Multiple of 50 Hz | 1 | 2 | (etc.) | | | | | |
| True frequency /Hz | 50 | 100 | (etc.) | | | | | |
| Scale error /Hz | | | | | | | | |

## GRAPH

Plot a graph of the scale error against the scale reading, so that it can be used to determine true frequencies.

# EXPERIMENT 55

## Measurement of Velocity of Sound in Free Air

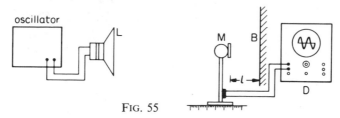

FIG. 55

### APPARATUS

A.F. oscillator, loudspeaker L with baffle, large board B, small microphone M connected to an oscilloscope D, thermometer 0–100 °C, metre rule.

### METHOD

Connect the oscillator to the loudspeaker and adjust the oscillator so that a sound of approximately 3000 Hz is produced by the speaker. Place the loudspeaker L facing the large board B, about 1 metre from it. Move the microphone M between L and B, along the axis of the loudspeaker, starting near B. Stand behind the board B to avoid stationary waves due to stray reflection from the experimenter. Looking at the oscilloscope screen, note the positions of the microphone when the trace has its smallest amplitude. These correspond to the positions of minimum sound, which are nodes of the standing waves in the space LB. Measure the length $l$ cm between M and board B for each node. Record the room temperature.

### MEASUREMENTS

|  | 1st node | 2nd node | 3rd node | 4th node |
|---|---|---|---|---|
| Length $l$/cm |  |  |  |  |
| Distance between alternate nodes, $\lambda$/cm | (3)−(1) |  | (4)−(2) |  |

Average value of $\lambda =$ .. cm $=$ .. m

### CALCULATION

Calculate the wavelength $\lambda$ by subtracting the readings of $l$ for alternate nodes, and enter them in the table of measurements. Then find the average value of $\lambda$. Calculate $V$ from $V=f\lambda$, where $f=$ frequency, $V=$ velocity of sound.

### CONCLUSION

The velocity of sound at room temperature ( .. °C) was found to be .. $\mathrm{m\,s^{-1}}$.

### ERRORS

1. The error in locating a node, or position of minimum sound, may be investigated by finding how far the microphone has to be moved to cause a just perceptible change amplitude of the trace on the oscilloscope screen. The error in $\lambda$ is equal to twice the error in each measurement of $l$, since pairs of values of $l$ are subtracted.

2. Error in the oscillator at a frequency $f$.

### ORDER OF ACCURACY

The percentage error in $V$ is given by:

$$\frac{\delta V}{V} \times 100\% = \left[ \frac{\delta f}{f} + \frac{2\delta l}{l} \right] \times 100\%, \text{ from } V=f\lambda$$

# PART SIX

# Electricity

# EXPERIMENT 56

## Investigation of Variation of P.D. of Cell and Measurement of Internal Resistance

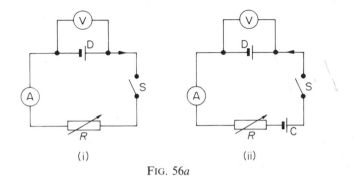

FIG. 56a

APPARATUS

1·5 V dry cell D, voltmeter V (0–2 V), ammeter A (about 0–0·5 A), rheostat $R$ (about 30 Ω max.), 2 V accumulator C, switch S.

METHOD

1. Connect the cell, ammeter A, rheostat $R$, and switch S in series, and the voltmeter V across the terminals of the cell, as shown in Fig. 56a (i). Set the rheostat at its maximum value, close the switch and read the current $I$ and the p.d. $V$. Vary the rheostat, and take about four pairs of readings of current and p.d. over as wide a range as possible, only closing the switch for as long as is necessary to take each reading, to avoid possible changes in e.m.f.

2. Connect the accumulator C, with its positive pole connected to that of the cell, as shown in Fig. 56a (ii). Reverse the connections to the ammeter (unless it is of a centre-zero type). Again setting the rheostat to its maximum value, read the current $I$ now being passed through the cell, and the p.d. developed across its terminals. Decrease the resistance, and again take about four pairs of readings over as wide a range as the meters allow. Record the values of current with a *minus sign*, as it is being passed in the opposite direction to that in part (i).

MEASUREMENTS

|  | (i) | | | | (ii) | | | |
|---|---|---|---|---|---|---|---|---|
| Current $I$/A | + | + | + | + | − | − | − | − |
| P.d. $V$ across cell /V | | | | | | | | |

136

GRAPH

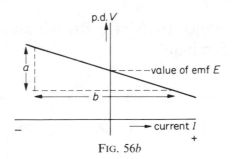

FIG. 56*b*

Plot the values of current, *I*, and p.d., *V*, as shown, taking the sign of the current in part (ii) to be negative. Since the p.d. *V* is given by $V = E - Ir$ when the cell supplies current, and by $V = E + Ir$ when current is 'forced' through the cell, then:

(i) The e.m.f. *E* is the value of *V* when $I = 0$.
(ii) The gradient of the line $= r$.

Record the value of *E*:     $E = $ .. V
Measure the gradient: $r = a/b = $ .. $\Omega$

CONCLUSIONS

1. The e.m.f. of the cell was .. V.
2. The internal resistance of the cell was .. $\Omega$.

ERRORS

1. The accuracy of the readings depends on the range and sensitivity of the meters used.
2. Draw another straight line roughly parallel with the 'best' line, such that it passes near or through the points on the graph which lie to one side of the 'best' line. Read off the new value of the e.m.f., and compare it with the 'best' value.
3. Draw another straight line of greater or of less slope than the 'best' line, passing through the same value of the e.m.f., and still agreeing fairly well with the points plotted. Re-measure the value of the slope, and compare it with the 'best' value of the internal resistance.

ORDER OF ACCURACY

Use the methods of (2) and (3) to estimate the maximum errors in *E* and *r*, and hence the order of accuracy of the final result.

NOTE

The internal resistance of a cell changes as it is used. Investigate this change.

# EXPERIMENT 57

## Measurement of Resistance by Metre Bridge and Investigation of Series and Parallel Formulae

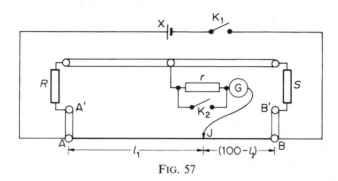

FIG. 57

### APPARATUS

Metre bridge, dry cell X, key $K_1$, galvanometer G, protective resistance $r$ and shorting key $K_2$, jockey J, two unknown resistances $R_1$, $R_2$, standard resistance S of the same order of magnitude as $R_1$, $R_2$.

### METHOD

Connect the cell X and the key $K_1$ to the terminals A, B at the ends of the resistance wire AB. Connect the galvanometer G, through the protective resistance $r$ (not less than $1000\,\Omega$), to the centre terminal D of the heavy brass strip CDE. Carefully clean the terminals C, E and A', B' with fine emery paper, and connect the unknown resistance $R_1$ and the known resistance S by short, cleaned, thick wires to C and A', and E and B' respectively, so that as little extra resistance as possible is introduced. The protective resistance $r$ is either a high (radio-type) resistance in series as shown, or a rheostat of a few ohms connected in parallel with the galvanometer.

Close the key $K_1$, and locate the balance point by touching the jockey J lightly onto the bridge wire. When the position of the balance point has been found to the nearest cm short the protective resistance by $K_2$ and locate the balance point as accurately as possible. Measure the length $l$ from the end A to the balance point.

Interchange the resistances $R_1$ and S, and repeat the measurement of $l$, which will now be measured from the end B. Repeat the experiment with: (i) the unknown resistances $R_2$; (ii) $R_1$ and $R_2$ connected in series; (iii) $R_1$ and $R_2$ connected in parallel.

### NOTE

The bridge is insensitive when the balance point is not near the centre of the wire. If necessary, the known resistance S should be replaced by one which gives a balance point between 30 and 70 cm of the wire.

MEASUREMENTS

| Unknown resistance | Known resistance $S/\Omega$ | Length $l$ /cm | Length $(100-l)$ /cm |
|---|---|---|---|
| $R_1$ | | } Average | } Average |
| $R_2$ | | } Average | } Average |
| $R_1, R_2$ in series | | } Average | } Average |
| $R_1, R_2$ in parallel | | } Average | } Average |

CALCULATION

By the usual Wheatstone bridge relation we have:

$$\frac{R_1}{S}=\frac{l_1}{100-l_1} \quad \text{or} \quad R_1=\frac{l_1}{(100-l_1)}.S= \ .. \ \Omega$$

Similarly,
$$R_2= \ .. \ \Omega$$
$$R_1, R_2 \ (\text{series})= \ .. \ \Omega$$
$$R_1, R_2 \ (\text{parallel})= \ .. \ \Omega$$
$$Series \ R=R_1+R_2= \ .. \ \Omega, \text{ by calculation}$$
$$Parallel \ 1/R=1/R_1+1/R_2= \ ..$$
$$\therefore \ R= \ .. \ \Omega, \text{ by calculation}$$

CONCLUSIONS

1. The resistances measured were .. $\Omega$.
2. The combined resistance of $R_1$ and $R_2$ is given, within the limits of experimental error, by the formula .., for series connection, and by the formula .., for parallel connection.

ERRORS

1. The length $l$, that is, the position of the jockey, may be read to, say, $\pm 1$ mm.
2. The accuracy with which the balance point may be *located* is also important. To investigate this, find the distance through which the jockey has to be moved to cause a *just* perceptible deflection of the galvanometer.
3. The bridge wire may not be exactly 1 metre long, so record its exact length and use this in the calculation.

ORDER OF ACCURACY

$$\text{Max. \% error in } R=\frac{\delta R}{R}\times 100\%=\left[\frac{\delta l}{l}+\frac{\delta(100-l)}{l}\right]\times 100\%$$

# EXPERIMENT 58

# Measurement of Resistivity of a Metal by Metre Bridge

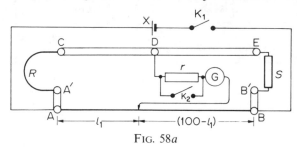

FIG. 58a

## APPARATUS

Metre bridge, dry cell X, key $K_1$, galvanometer G with protective resistance $r$ and shorting key $K_2$ not less than $1\,k\Omega$, jockey J, 2–3 m of nichrome or constantan wire of s.w.g. 22–24, suitable standard resistances S, micrometer gauge, metre rule.

## METHOD

1. Connect the cell X and key $K_1$ to the terminals A, B of the wire AB (Fig. 58a). Connect the galvano-meter G to the jockey J and D, using the protective resistance $r$. The protective resistance $r$ may be a high resistance in series with G and a shorting key $K_2$ as shown, or a rheostat of a few ohms in parallel with G. Carefully clean the terminals C, E and A′, B′ with fine emery paper and connect the standard resistance S between E and B′.

2. Remove any kinks from the coil of wire R whose resistivity is required and connect the whole length of it to the terminals C and A′. Make a sharp bend at one end where it leaves the terminal C say and an ink mark where it leaves the other terminal A′.

Close the key $K_1$ and locate the balance point by touching J lightly on the bridge wire. When the position of the balance point has been found to the nearest centimetre, remove the protective resistance $r$ by closing $K_2$ and locate the balance point as accurately as possible. Measure the length $l_1$ from the end A to the balance point. (If necessary, change the standard resistance S so that the balance point is nearer the middle of the wire.) Now measure carefully the length $l$ of wire between the terminals C and A′.

3. Repeat the experiment with four shorter lengths of the wire and record the values of $l_1$ and $l$. You may have to change the standard resistance to a smaller value in order to obtain a balance point near the middle of AB.

4. Interchange the resistance S and the wire R and repeat the measurements of $l_1$ (this time measured from B) for each length $l$ of wire used previously. Enter the results in the table so that an average value for $l_1$ can be found.

5. Remove the wire and measure its diameter at three different places with a micrometer gauge; at each place take two measurements at right angles. Record also the zero error, if any, of the micro-meter gauge.

## MEASUREMENTS

| Length of resistance wire $l/m$ | .. |
| --- | --- |
| Resistance $S/\Omega$ | .. |
| Length $l_1/mm$ | .., .. Av.= .. |
| Length $(100-l_1)/mm$ | .., .. Av.= .. |

Diameter of wire $= \ldots, \ldots, \ldots$ mm
Average diameter $= \ldots$ mm
Zero error of micrometer $= \ldots$ mm
Corrected diameter $d = \ldots$ mm $= \ldots \times 10^{-3}$ m

140

## CALCULATION AND GRAPH

Calculate the resistance $R$ for each length $l$ of the wire from

$$R = \frac{l_1}{100 - l_1} \times S$$

Enter the results for $R$ in the table.

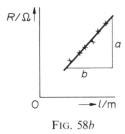

FIG. 58b

Plot a graph of $R$ in ohm against $l$ in metre (Fig. 58b).
Measure the gradient $a/b$ of the best straight line through the points. Now

$$R = \frac{\rho l}{A} = \frac{\rho l}{\pi d^2/4} = \frac{4\rho l}{\pi d^2}$$

So

$$\text{gradient} = \frac{4\rho}{\pi d^2}$$

Calculate the resistivity from

$$\rho = \frac{\pi d^2}{4} \times \text{gradient} \quad (d \text{ in metre})$$

## CONCLUSION

The resistivity of the metal ... = ... $\Omega$ m.

## ERRORS. ORDER OF ACCURACY

The most serious error may occur in the measurement of the length $l$ of wire actually emerging from the terminals. There is also uncertainty in the measurement of the balance length $l_1$—determine the distance on the wire through which the jockey has to be moved to produce a *just* perceptible deflection in the galvanometer. The metre bridge wire may not be exactly 1 metre long, too.

The errors in the measurement of the diameter $d$ arise from: (i) lack of uniformity in the wire; (ii) random errors of observation in the micrometer, both of the actual diameter and of the zero reading. Since the diameter $d$ is squared, the error in this will be doubled, and may well be the most significant error in the experiment (see p. 8).

The order of accuracy can be determined by the method of drawing the lines of least and greatest gradients which pass through the points on the graph (p. 9).

# EXPERIMENT 59

# Measurement of Temperature Coefficient of Resistance

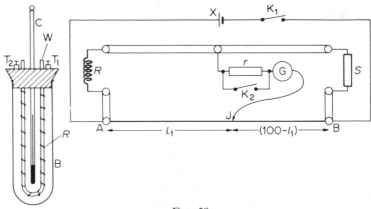

FIG. 59a

## APPARATUS

Metre bridge, dry cell X, key $K_1$, galvanometer G with protective high series resistance $r$ and shorting key $K_2$, jockey J, coil of iron wire R, thermometer 0–100 °C, C, large beaker, bunsen, tripod, gauze, ice, stirrer, standard resistance S.

## METHOD

1. Connect up the metre bridge circuit for measuring the resistance R of the coil W whose temperature coefficient of resistance is required. For details, see Expt. 57. S is a suitable standard resistance, about the same resistance as W at room temperature.

2. Immerse the coil W in crushed, melting ice in a large beaker (see Note 1 if ice is not available). Wait until the temperature of W is steady at 0 °C and determine the balance point on the bridge wire AB. Record the balance length $l$ at this temperature.

3. Warm the ice until it melts, add water to the beaker, and measure the new balance length when the temperature has risen about 10 K and is steady. Record the exact value of the temperature. Continue to warm the water further, maintain the temperature steady at a series of temperatures up to about 90 °C, and record the balance length at each temperature.

4. Allow the water to cool, adding cold water if necessary, and repeat the measurement of the balance lengths at the temperatures previously recorded so that an average value can be found.

## MEASUREMENTS

| Temperature $t$ /°C | Length $l$ /cm | length $(100 - l)$ /cm | resistance S /Ω | resistance $R_t$ /Ω |
|---|---|---|---|---|
| .. | .., .. Av. = .. | .., .. Av. = .. | .. | .. |

142

## CALCULATION. GRAPH

Calculate the resistance $R_t$ at each temperature $t$ in °C from the relation

$$R_t = \frac{l_1}{100 - l_1} \times S$$

Plot a graph of the resistance $R_t$ v. temperature $t$, and draw the best straight line, as shown (Fig. 59b).

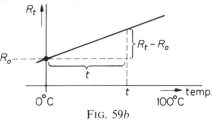

FIG. 59b

The temperature coefficient of resistance $\alpha = \dfrac{R_t - R_0}{R_0 \times t}$.

But
$$\frac{R_t - R_0}{t} = \text{gradient of graph}$$

$$\therefore \ \alpha = \frac{\text{gradient of graph}}{R_0} = \ ..$$

## CONCLUSION

The temperature coefficient of resistance of $..$ = $..$ K$^{-1}$.

## NOTES

1. If ice is not available, produce the graph back to the $R_t$ axis to find $R_0$, the resistance at $0\,°C$.

2. A suitable resistance coil may be constructed by winding a length of thin iron wire on a bent glass rod, soldering the ends to terminals $T_1$, $T_2$, secured in a cork (see Fig. 59a). A hole in the centre of the cork takes a thermometer C, and the whole is enclosed in a boiling tube B, which may be filled with a light oil to improve the thermal contact. Alternatively, the coil should be well varnished, when it may be immersed directly in a water bath. Thin copper wire, preferably enamelled, may also be used.

## ERRORS

1. See p. 141 for errors in the metre bridge experiment.

2. The thermometer may be read to, say, $\pm 0\cdot 1\,K$ (a much larger percentage error).

3. The coil may not be at the temperature indicated by the thermometer.

4. The gradient of the graph can only be found to an accuracy limited by the scale on which it is plotted.

## ORDER OF ACCURACY

Draw the line of greatest or least slope, which differs from that of the 'best' line, and re-measure the gradient. The percentage error in the temperature coefficient is the sum of the percentage errors in the gradient and in $R_0$. Hence determine the order of accuracy.

# EXPERIMENT 60

## Comparison of E.m.f.s of Cells by Potentiometer

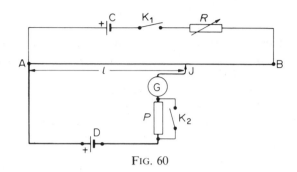

FIG. 60

APPARATUS

Dry cell D, potentiometer, key $K_1$, galvanometer G with protective resistance $P$ and shorting key $K_2$, jockey J, accumulator C, Weston or other standard cell, rheostat $R$.

METHOD

Connect the accumulator C to the ends A, B of the potentiometer wire and rheostat $R$, which should initially be set at zero resistance. Connect the positive terminal of the cell D to the same end A as that to which the positive terminal of C is connected. Join the negative terminal of D through the protective resistance $P$ and the galvanometer G to the jockey J.

Touching the jockey lightly on the potentiometer wire, find the position of the balance point to within the nearest centimetre. Remove the protective resistance by the shorting key $K_2$ and measure the distance $l$ of the balance point from the end A as accurately as possible. Repeat the setting several times and take the average value of $l$. Now replace the cell D by a Weston standard cell with a large resistance in series, again with its positive pole connected to A, and measure the length $l_W$ required for a balance, as above.

Increase the resistance in the rheostat $R$, thus decreasing the current and the p.d. per cm in AB, and repeat the measurements for $l$ and $l_W$ for, say, two different settings of the rheostat.

MEASUREMENTS

| Various rheostat settings | 1 | 2 | 3 |
|---|---|---|---|
| Length $l$ required for balance with cell /m | | | |
| Length $l_W$    ,,      ,,      ,,      ,,    Weston cell /m | | | |

E.m.f. of Weston cell at room temperature = .. V

CALCULATION

Expt. (1) $E = (l/l_W) \times E_W = $ .. V
Expt. (2) $E = $          $= $ .. V
Expt. (3) $E = $          $= $ .. V
Average $E = $        $= $ .. V

144

## CONCLUSION

The e.m.f. of the cell was found to be .. V.

## ERRORS

1. The theory assumes that there is a steady current in AB while any set of measurements is taken. Any tendency of the balance point to 'drift' towards the end B should be watched for, as it indicates that the accumulator is running down.

2. Measure how far the jockey must be moved to cause a just perceptible deflection on the galvanometer, to estimate the error in locating the balance.

3. Note the accuracy with which the position of the jockey can be read from the scale, say $\pm 1$ mm. There is also a zero error in the measurement of $l$, due to the end of the scale not being exactly at the end of the wire.

## ORDER OF ACCURACY

From the formula for $E$,

$$\text{max. \% error in } E = \frac{\delta E}{E} \times 100\% = \left( \frac{\delta l}{l} + \frac{\delta l_w}{l_w} \right) \times 100\%$$

## NOTE

Investigate how quickly the e.m.f. of the dry cell recovers after it is short-circuited.

# EXPERIMENT 61

## Calibration of Ammeter by Potentiometer

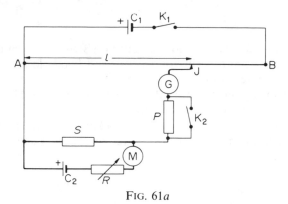

FIG. 61a

### APPARATUS

Potentiometer, key $K_1$, two accumulators $C_1$, $C_2$, rheostat $R$, ammeter M (about 0–2 A), standard resistance $S$ (1 Ω), galvanometer G and protective resistance $P$ with shorting key $K_2$, jockey J, Weston (standard) cell.

### METHOD

Connect the accumulator $C_1$ to the potentiometer wire AB and key $K_1$. Connect the rheostat $R$, ammeter M, and accumulator $C_2$ in series with the standard resistance $S$. Join the terminal of $S$ which is connected to the positive pole of $C_2$ to the end A of the balance wire; this is also joined to the positive pole of $C_1$. Join the other terminal of $S$ in series with the galvanometer G, protective resistance $P$, and jockey J. The terminals of $S$ should be freshly cleaned.

Set the ammeter M at a reading such as 0·20 A by means of $R$, and obtain the balance point to the nearest cm; then remove the protective resistance by the shorting key $K_2$ and obtain the final balance to within 1 mm. Set the ammeter at 0·40 A, and repeat, having replaced the protective resistance. A current in excess of 2 A should not be used, as this may heat $S$ and so affect its resistance appreciably. Now remove the ammeter circuit ($C_2$, $R$, M, $S$) completely. Measure the length $l_W$ of wire which 'balances' a Weston cell, using a protective resistance.

### MEASUREMENTS

| Current indicated on ammeter /A | | | | |
|---|---|---|---|---|
| Length of wire /m | | | | |
| Current (true) /A | | | | |
| Correction /A | | | | |

Length $l_W$ corresponding to e.m.f. of Weston cell = .. m
E.m.f. $E_W$ of Weston cell at room temperature = .. V

146

## CALCULATION

The p.d. $V$ across $S$ is given by

$$\frac{V}{E_w} = \frac{l}{l_w} \quad \text{or} \quad V = \frac{l}{l_w} \times E_w$$

$$\therefore \quad \text{true current for } 0.20 \text{ A reading} = \frac{V}{S} = \text{ .. A} \quad . \quad . \quad . \quad . \quad . \quad . \quad \text{(i)}$$

$$\text{,,} \qquad \text{,,} \qquad \text{,, } 0.40 \text{ ,,} \qquad \text{,,} \qquad = \frac{V}{S} = \text{ .. A} \quad . \quad . \quad . \quad . \quad . \quad . \quad \text{(ii)}$$

$$\text{etc.}$$

Calculate the *correction* at each point of the ammeter scale, i.e. the amount which needs to be added to the indicated current to give the true current.

## GRAPH

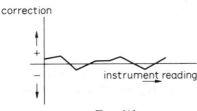

FIG. 61*b*

Draw a graph of 'correction v. instrument reading'. Join the points by a series of straight lines if the errors bear no relation to each other (Fig. 61*b*).

## CONCLUSION

The correction to be added to the indicated current is shown in the graph.

## ERRORS

1. This experiment is valid only if the errors of measurement, and of setting the ammeter readings, are considerably smaller than the actual errors present in the instrument. If this is impossible to achieve, the ammeter should be shunted with a resistance sufficient to introduce an error of some 5%.

2. Errors occur in the measurement of $l/l_w$, see p. 145, Expt. 60.

3. If there is poor contact at the terminals of the standard resistance the values of 'true current' will be systematically low.

## ORDER OF ACCURACY

See p. 145.

# EXPERIMENT 62

# Measurement of Internal Resistance of Cell by Potentiometer

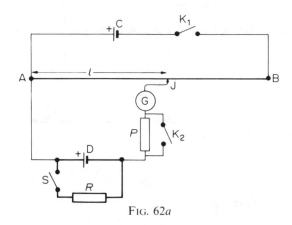

FIG. 62a

## APPARATUS

Dry cell D, resistance box $R$ (about 0–50 $\Omega$), switch S, potentiometer, key $K_1$, galvanometer G and protective resistance $P$ with shorting key $K_2$, jockey J, accumulator C.

## METHOD

Connect the accumulator C to the ends A, B of the potentiometer wire. Connect the positive pole of the cell D to the same end, A, connected to the positive pole of C. Join the galvanometer G and its protective resistance $P$ to the jockey J and to the negative pole of D. Connect the resistance box $R$ and switch S across the terminals of the cell D.

With the switch S open, find to the nearest cm the position of the balance point, removing the protective resistance to obtain the final balance length $l_0$ corresponding to $E$, the e.m.f. of cell D; this is the terminal p.d. when it passes no current. Then, with $R$ set at about 10 $\Omega$, close S and quickly measure the position of the new balance point. Open S as soon as this has been done. Reduce the value of $R$ in roughly five equal steps to 1 $\Omega$, and each time obtain the balance length as quickly as possible. At the end of the experiment, keep the switch S open, and re-measure the open-circuit balance length $l_0$.

## MEASUREMENTS

Length of wire for cell D on open circuit $= l_0 = $ .. m

| $R/\Omega$ | | | | | | | | | |
|---|---|---|---|---|---|---|---|---|---|
| $l/m$ | | | | | | | | | |
| $(1/R)/\Omega^{-1}$ | | | | | | | | | |
| $l_0/l$ | | | | | | | | | |

# CALCULATION

Internal resistance $r = \dfrac{E-V}{I} = \dfrac{E-V}{V/R} = \left(\dfrac{E-V}{V}\right)R$

$$\therefore\; r = \left(\dfrac{l_0-l}{l}\right)R = \left(\dfrac{l_0}{l}-1\right)R. \quad \cdot \quad \cdot \quad \cdot \quad \cdot \quad \cdot \quad \cdot \quad \cdot \quad \cdot \quad \cdot \quad (1)$$

or

$$\dfrac{l_0}{l} = r\left(\dfrac{1}{R}\right)+1 \quad \cdot \quad \cdot \quad \cdot \quad \cdot \quad \cdot \quad \cdot \quad \cdot \quad \cdot \quad \cdot \quad \cdot \quad (2)$$

# GRAPH

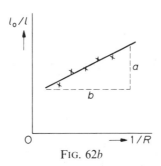

FIG. 62*b*

Plot a graph of $(l_0/l)$ v. $1/R$. From equation (2), this will be a straight line of gradient $r$. Measure the gradient, $a/b$, of the best straight line drawn through all the points (Fig. 62*b*).

$$r = \dfrac{a}{b} = \; \dots \; \Omega$$

# CONCLUSION

The internal resistance was .. $\Omega$.

# ERRORS

1. For errors arising in the measurement of a balance length see Expt. 60, p. 145.
2. The cell may polarise during the experiment; this may cause the graph to be appreciably curved where $1/R$ is large.

# ORDER OF ACCURACY

Estimate the order of accuracy by drawing lines which have slopes differing as widely as possible from the 'best' line, while just agreeing with most of the plotted points.

# NOTE

From (2), when $l_0/l = 0$, then $(1/r) = -(1/R)$. Thus $r$ may be found from the negative intercept on the $1/R$-axis when the straight line is produced back. However, with a battery of low resistance, say less than 1 ohm, the intercept would be subject to appreciable error depending on the slope of the straight line drawn through the points.

# EXPERIMENT 63

## Measurement of Thermoelectric E.m.f. and Variation with Temperature

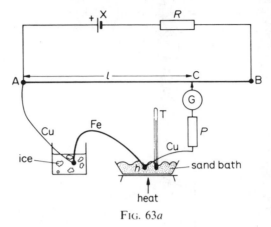

FIG. 63a

APPARATUS

2 V accumulator X, copper (Cu)–iron (Fe) thermocouple, variable resistance box $R$ up to several thousand ohms, potentiometer AB, key $K_1$, very sensitive galvanometer G and protective resistor $P$ with shorting key, jockey or slider C, ice, beaker, sand bath, thermometer 0–350 °C, tripod, gauze, bunsen burner, metre bridge and accessories, and a good voltmeter.

METHOD

Measure the resistance $S$ of the wire AB using a metre bridge (see p. 138). Measure the e.m.f. $E$ of the accumulator X using a good voltmeter. Connect up the potentiometer circuit with the thermocouple as shown in the diagram, with $R$ set temporarily at about 500 Ω, in which case the p.d. across AB, if it has 1 Ω resistance, is 4 mV. Note that the cold junction $c$, surrounded by melting ice in a beaker, is joined to A, to which the *positive* terminal of X is connected. Warm the other junction $h$ in the fingers and check that a balance is obtainable close to A. Place this 'hot' junction $h$ in the sand bath close to the bulb of the thermometer, and heat the bath gently. As the temperature rises, locate the approximate balance points, until the balance length $l$ reaches a maximum near 300 °C. Maintain this temperature fairly steady with a small flame, and increase the resistance $R$ until the balance length $l$ is about 75 cm of the 100 cm wire AB. Write down this final value of $R$, and do not alter it further.

Now heat the sand bath to 350 °C, and allow it to cool slowly to room temperature, taking careful readings of the balance length $l$ after removing the protective resistor, every 20 K. The cooling of the bath may be retarded by using a small flame.

MEASUREMENTS

Resistance of wire AB, $S$   = .. Ω
Resistance, $R$   = .. Ω
Cold junction temperature = 0 °C

| Hot junction temperature $t$/°C | Balance length $l$/m | E.m.f. of thermo-couple /mV |
|---|---|---|
|  |  |  |

150

## CALCULATION

The p.d. in millivolt across the wire AB, 100 cm long, resistance $S$, is given by:

$$\text{p.d. across AB} = \frac{1000E}{R+S} \times S$$

where $E$ is the e.m.f. of the accumulator in volt. Calculate the thermo-electric e.m.f. in millivolts for each temperature from the equation:

$$\text{thermo-electric e.m.f.} = \frac{l}{100} \times \text{p.d. across AB} = \frac{l}{100} \times \frac{1000ES}{(R+S)}$$

Enter the values of the e.m.f. in the table of measurements.

## GRAPH

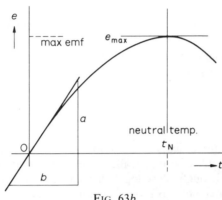

FIG. 63b

Plot a graph of the thermo-electric e.m.f. $e$ v. temperature of the hot junction, $t\,°C$ (Fig. 63b).
 Find the *maximum e.m.f.*, $e_{max.}$, reached, and the *neutral temperature*, $t_N$, at which this occurs.
 Measure the slope $a/b$ of the graph at the origin O, which is equal to the constant $A$ in the equation:

$$e = At + Bt^2$$

which represents the behaviour of most thermocouples. Since $de/dt = 0$ at the maximum value of $e$, the value of $B$ may be found from the equation $A + 2Bt_N = 0$, obtained by differentiating the formula for $e$.

## CONCLUSIONS

The maximum e.m.f., $e_{max.}$, was found to be .. mV.
The neutral temperature $t_N$ was found to be .. °C.
The constant $A$ was found to be .. $mV\,K^{-1}$.
The constant $B$ was found to be .. $mV\,K^{-2}$.

## ERRORS

These occur in measuring the balance length $l$, the e.m.f. $E$, and the resistance $S$ (see p. 161).

# EXPERIMENT 64

## Absolute Measurement of Current

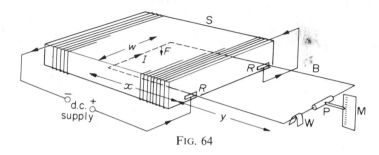

FIG. 64

APPARATUS

Flat solenoid S (as shown, e.g. Griffin & George, L83–740) or conventional solenoid (e.g. Griffin & George, L83–750), either about 0·3 m long, with about 300 turns or more; rectangular wire frame current balance B, to fit inside solenoid, with mounted razor blades R to act as pivots and electrical connections; paper tape for weights W; d.c. supply 0–12 V; rheostat about 10 Ω to carry at least 5 A; ammeter 0–10 A; leads.

METHOD

1. *Principle.* One ampere is defined (leaving out some refinements) as the current which produces a force of $2 \times 10^{-7}$ N on each metre of long parallel wires a metre apart, when they both carry this current. Even when the wires are brought close together the force is still small, and difficult to measure.

To increase the force, it is convenient to use a solenoid instead of one of the wires. The force on a wire at the centre of the solenoid can be calculated from the definition of the ampere, so this can be used to measure a current without using any other ammeter. The force is still small, but is now easier to measure (a precise measurement would normally use a still more complicated arrangement).

So the experiment consists of passing the *same* current through a solenoid and through a wire inside the solenoid, measuring the force $F$ on the wire, and calculating the current from $F$.

2. *The solenoid.* To make the force larger, by having a longer wire in the solenoid, the flat-shaped solenoid shown has advantages but a narrower one will also serve. Connect the solenoid to the d.c. supply, with a rheostat in series and with an ammeter temporarily in the circuit. Adjust the supply and the rheostat until the current is at least 5 A, preferably more but not so large as to overheat the solenoid. Remove the ammeter, which is not needed again, as this is an absolute measurement.

3. *The current balance.* The current balance B is made from stiff copper wire bent into a rectangular frame which just slides into the solenoid, as shown. Current is led into the frame via two mounted razor blades R near the end of the solenoid. The balance rests on the razor blades (nicks in the wire to locate it are useful). The balance should lie horizontally and swing freely. It is bent a little at the point of suspension so that it is stable. Current $I$ only passes in the part inside the coil, as shown: the arms outside are bridged by an insulating sleeve carrying a pin P. The pin moves near a scale M, which indicates the rest position of the balance.

Adjust the balance, using a little wire wrapped round one arm, until it is horizontal. Measure the distances $x$ and $y$ from the pivot to the two end arms of the balance and record the results. Position the mounted razor blades near the end of the solenoid, raising them a little if necessary until the balance lies at rest in the middle of the solenoid.

4. *Measurement of force.* Break one connection to the solenoid S, and put the current balance in series with the solenoid, as shown. Switch on, and note whether the wire in the solenoid is pushed down or up. If it is pushed up, reverse the connections to the current balance, so that it goes down.

152

Note the position of the pin P on the scale M for zero current. Now switch on the current, and then hang a few centimetres of paper tape (thin wire will do as well) as a counterweight W on the arm outside the balance. Cut pieces from the counterweight, or use a larger one, until the pin P is at the previous zero position when the current is flowing.

Note that the heat generated at the contact with the razor blades will soon damage them. This will make the current fluctuate and so make it difficult to obtain a consistent balance. If necessary, replace the blades.

(a) Weigh the counterpoise on an accurate balance and obtain its mass $m$ in kg.

(b) Count the number of turns $N$ in a length $L$ of the solenoid. ($L$ ought to be at least $0.1$ m to obtain an accurate estimate of the ratio $N/L$. Why?)

(c) Measure the width $w$ of the wire inside the balance (see Fig. 64).

## MEASUREMENTS

*Force*            Distance $x$ from pivot to wire           $= \,.\, .$ m
                   Distance $y$ from pivot to counterpoise $= \,.\, .$ m
                   Mass $m$ of counterpoise                $= \,.\, .$ kg
                   Width $w$ of wire                       $= \,.\, .$ m

*Solenoid*         Length $L = \,.\, .$ m; number of turns $N = \,.\, .$

## CALCULATION

As a first step, a calculation can be made without allowing for failure of ideal conditions.

Taking moments about the axis RR, the force $F$ on the wire is

$$F = \frac{mgy}{x} = \,.\, .\ N$$

If the solenoid were infinitely long and carried a current $I$, the flux density $B$ in it would be

$$B = \frac{\mu_0 N I}{l} = 4\pi \times 10^{-7} \frac{NI}{l}$$

The force $F$ on the wire of length $l = w$ is also given by $F = BIl = BIw$

Substituting for $B$, we obtain           $$F = 4\pi \times 10^{-7} \frac{NwI^2}{L}$$

So           $$I = \sqrt{\frac{FL}{4\pi \times 10^{-7} Nw}} = \,.\, .\ A$$

## DISCUSSION

1. *Other related absolute values.* The current is not the only quantity you have in effect obtained absolutely. The flux density $B$ can now be found, substituting $I$ in the equation for $B$. The flux $\Phi$ in the solenoid can then be found by multiplying $B$ by the cross-sectional area $A$ of the solenoid. In principle, voltages can then be measured using rate of change of flux.

2. *Effect of a short solenoid.* The field $B$ in the solenoid will be less than the value given by the equation for $B$ given above, because the solenoid has a finite length. If no allowance is made for finite length, the calculated value of $I$ can be as high as $1.5$ times the actual value.

3. *Other errors.* Consider what change in the mass $m$ of the counterpoise would lead to a detectable movement of the balance (try it). The force, and so the value of $I^2$, cannot be known with greater precision. Discuss whether this is the most serious source of error, or not.

# EXPERIMENT 65

# Investigation of Laws of Electromagnetic Induction

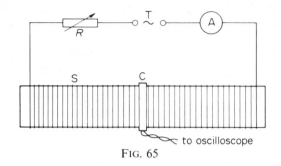

FIG. 65

## APPARATUS

Long solenoid S, about 0·3 m long, with about 300 turns; insulated wire to make coils C around the solenoid; transformer T 0–12 V; rheostat $R$, 15 Ω, 5 A; a.c. ammeter A, 0–5 A; oscilloscope; signal generator; multirange meter.

## METHOD

### Arrangement and Circuit

The solenoid S is connected in series to the rheostat $R$, transformer T, and a.c. ammeter A. Note that the rheostat is itself a solenoid, so place it well away from the other apparatus.

Coils C will be wound from insulating wire, to fit around or inside the solenoid. They can be wound round wooden blocks as formers, and kept in shape using adhesive tape. The alternating voltage across each coil, due to the changing flux through it, will be indicated on an oscilloscope connected to the coil. Generally, we investigate how the e.m.f. or voltage $E$ induced across a coil is related to the rate of change of the flux linkages $\Phi$ of the coil. If the magnetic flux density $B$ is uniform across an area $A$ of each turn, and the flux links each turn normally, then

$$\Phi = NAB$$

where $N$ is the number of turns of the coil.

### 1. Induced E.m.f. and Number of Turns

Wrap two turns of insulated wire to make a coil around the middle of the solenoid. Connect the coil to the oscilloscope, and increase the current in the solenoid, and the sensitivity of the oscilloscope, until a trace perhaps 10 mm from peak to peak is obtained. Switch off the time base, and measure the peak to peak amplitude of the trace, $a$.

Leaving the current and sensitivity the same, wrap more turns to make a coil of 4, 6, 8, 10, etc. turns. If the oscilloscope is not very sensitive, it may be better to start with 5 turns, and increase the number in fives. For each number of turns $N$, record the trace amplitude $a$ in a Table.

### 2. Induced E.m.f. and Area

Make two 20 turn coils, both of which will fit *inside* the solenoid, but each having a different area. (One wound round the outside of a match-box, and one wound round its narrowest cross-section, are convenient.)

Push each coil in turn into the centre of the solenoid, fixing them so that their axes are along the central axis of the solenoid. Using a suitable sensitivity on the oscilloscope, record the trace amplitude obtained from each coil in turn. Measure the width $w$ and breadth $b$ of each coil.

154

Check, without detailed measurement, that it is the flux passing through the area of the coil that matters, by turning one coil gradually until it is edge-on to the axis of the solenoid. As you do so, the indication on the oscilloscope may be expected to fall, reaching zero.

Also, make a coil of cross-section larger than the solenoid, and place it loosely round the middle. Investigate whether the oscilloscope indication is practically the *same* as with a coil having the same number of turns wound tightly round the solenoid. If so, what do you deduce?

## 3. Induced E.m.f. and Rate of Change of Flux

Replace the transformer by a signal generator having a low impedance output, if available. Remove the rheostat (use the gain control on the generator instead in order to vary the a.c. input), and replace the a.c. ammeter by a multirange a.c. ammeter.

Using a 10 or 20 turn coil wound round the middle of the solenoid, vary the frequency of the alternating current from the signal generator. Use the gain control to keep the current, monitored on the multimeter, constant. A frequency range from 50 Hz to 500 Hz may be suitable. Find a suitable oscilloscope sensitivity, and record the trace amplitude $a$ at a number of different frequencies $f$ in a Table.

MEASUREMENTS

| Turns $N$ | | | | | |
|---|---|---|---|---|---|
| Trace amplitude $a$ /mm | | | | | |

| Area $A = w.b$ /m$^2$ | (coil 1) | (coil 2) |
|---|---|---|
| Trace amplitude $a$ /mm | | |

| Frequency $f$ /Hz | | | | |
|---|---|---|---|---|
| Trace amplitude $a$ /mm | | | | |

GRAPHS AND CALCULATION

1. Plot a graph of trace amplitude $a$, which is proportional to the maximum induced voltage, against turns $N$. Is a straight-line graph obtained which passes through the origin?

2. Calculate the ratio of the areas of the two small coils $w_1 b_1$ and $w_2 b_2$; calculate the ratio of the corresponding trace amplitudes $a_1$ and $a_2$. Are these two ratios the same? A complicating factor will arise if the wire is thick and occupies an appreciable part of the cross-sectional area. Consider whether this might explain any discrepancy in the two ratios.

3. If the current in the solenoid, and hence the flux, varies sinusoidally, so that

$$\Phi = \Phi_0 \sin 2\pi ft$$

then

$$d\Phi/dt = 2\pi f \Phi_0 \cos 2\pi ft$$

So the maximum value of the rate of change of flux, for the same maximum flux $\Phi_0$, will be proportional to $f$. Plot a graph of the trace amplitude $a$ against $f$. Is this a straight-line graph passing through the origin?

CONCLUSIONS

Discuss whether your results are in agreement with the theory that the induced e.m.f. in a coil is proportional to the rate of change of flux linkages and whether any discrepancies can be explained by defects in the experimental arrangement.

# EXPERIMENT 66

# Investigation of Field of Solenoid with Search Coil

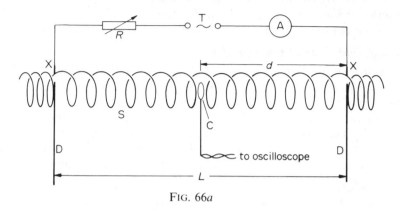

FIG. 66a

## APPARATUS

Large Slinky spring S to be used as a solenoid (alternatively, set of solenoids such as Griffin & George L 83–750); 5000 turn search coil C (e.g. Griffin & George, L 83–772); oscilloscope; low-voltage transformer T giving up to 12 V; rheostat $R$, 0–15 $\Omega$, 5 A; a.c. ammeter A, 0–5 A; metre rule; crocodile clips X; leads; strips of wood B to hold the spring stretched.

## METHOD

### 1. Arrangement and Circuit

Stretch out a part of the spring evenly along the bench, about 0·5 m long and with the turns about 10 mm apart, using strips of wood D held in clamps. Connect across this part of the spring, using crocodile clips, the transformer T, rheostat $R$ and a.c. ammeter A in series. The rheostat is also a solenoid and should be placed well away from the spring.

Connect the search coil C to the oscilloscope, with the oscilloscope set initially at its greatest sensitivity (for example, 0·1 V cm$^{-1}$). Push C sideways into the solenoid as shown, with its axis along that of the solenoid.

Pass an alternating current of about 1 A in the solenoid, and check that a trace can be obtained on the oscilloscope with the search coil in the middle of the solenoid. Increase the current, or alter the oscilloscope sensitivity if necessary, in order to obtain a trace which occupies about half the height of the screen. Use the time base and trigger controls to obtain a steady trace; or switch off the time base so that the trace is a vertical line.

The alternating voltage across the search coil is proportional to the rate of change of flux through it. Since the frequency is fixed (50 Hz), the amplitude of the voltage is proportional to the maximum flux. Since the area and number of turns of the search coil are constant, this is proportional to the flux density $B$ at the search coil.

### 2. Investigation of Flux Density B and Current I

Make a series of observations of the peak to peak amplitude $a$ of the oscilloscope trace for different currents $I$ in the solenoid, with the search coil fixed in one position.

156

### 3. Investigation of Effect of Turns per Unit length

Move the supports D closer together or further apart, so as to stretch the solenoid to various lengths $L$, between the points where the connections X are made. On each occasion record $L$ and the number of turns $N$ in this distance, and, keeping the current constant, observe the amplitude $a$ of the oscilloscope trace.

### 4. Variation of Field along the Axis

Put the search coil C into the solenoid at various points along its axis, noting its distance $d$ from one end. Observe the amplitude $a$ of the trace, at each position. (Continue the observations beyond the end of the current-carrying part of the solenoid.)

### 5. Other Variations (qualitative)

(a) Push the search coil across the width of the solenoid, and note that the flux density is very nearly constant across the whole width, falling sharply to zero as the search coil passes out of the solenoid.

(b) Move the connections X closer together, with the search coil at the centre of the solenoid, and the current and turns density kept the same (adjust the rheostat to allow for the changing resistance when current passes in less of the spring). Note that the flux density at the centre remains very nearly constant, until the distance between the connections is of the same order as the diameter of the solenoid, when it begins to fall.

(*Experiment continued overleaf*)

# 66. (*Continued*)

## MEASUREMENTS

| Current $I$/A | | | | |
|---|---|---|---|---|
| Trace amplitude $a$/mm | | | | |

| Length $L$/m | | | | |
|---|---|---|---|---|
| Turns density $(N/L)\,\mathrm{m}^{-1}$ | | | | |
| Trace amplitude $a$/mm | | | | |

| Distance $d$/m | | | |
|---|---|---|---|
| Trace amplitude $a$/mm | | | |

## GRAPHS

1. Plot a graph of trace amplitude $a$, which is proportional to the flux density $B$, against the current $I$ in the solenoid. Is the graph a straight line through the origin?

2. The flux density $B$ in an infinitely long solenoid, in which $N$ turns occupy length $L$, is given by

$$B = \mu_0 \frac{NI}{L}$$

Plot a graph of the trace amplitude $a$ against the turns density $N/L$. The graph may be expected to be a straight line through the origin, except perhaps at large values of $N/L$, where $L$ is not much greater than the diameter of the solenoid (see 5(*b*) above).

3. Plot a graph of the variation of the trace amplitude $a$ against the distance $d$ from one end of a fixed solenoid. The graph may be similar to that shown in Fig. 66b. The equation of the curve can be found in textbooks. One of its predictions is that the flux density exactly at the end ($d=0$ or $d=L$) of the solenoid, is just half the value at the middle. Check this prediction.

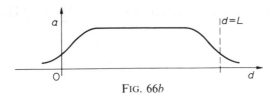

FIG. 66b

CONCLUSIONS

From your graphs, state what conclusions can be drawn on how $B$ varies (i) with $I$, (ii) with $N/L$ and (iii) with distance $d$ from one end.

QUESTIONS

1. The results depend on the height of the trace on the oscilloscope being proportional to the voltage applied to it. How would you check this?

2. Outline how you would combine knowledge of the sensitivity of the oscilloscope, the number of turns and area of the search coil, and the frequency of the supply, to convert the amplitude $a$ into the maximum flux density $B$, and so obtain $\mu_0$.

# EXPERIMENT 67
# Hall Effect. Measurement of $B$ inside Solenoid

## A. INVESTIGATION OF HALL EFFECT IN SEMICONDUCTOR

### APPARATUS

p- or n-semiconductor slice S, milliammeter A (0–100 mA), millivoltmeter G (0–30 mV, e.g. microammeter 0–100 μA, 1000 Ω, UNILAB, or 75–0–75 μA, 500 Ω WHITE), rheostat P (1 kΩ), 3 V battery D, Ticonal magnet M.

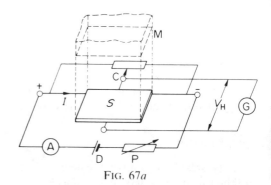

FIG. 67a

### METHOD

1. Connect the semiconductor S, say p-type, in series with D, P and A. Connect the meter G across the potentiometer terminal C and the opposite side of S, so that G will later measure the current or p.d. due to $V_H$, the Hall voltage developed across S *transverse* to the main current $I$ when the magnet M is brought up. Use the rheostat P to obtain a current $I$ of a few milliamperes. $I$ must not exceed the maximum stated by the supplier, or the semiconductor may be permanently damaged.

2. *Initial adjustment.* It is difficult to obtain two points on S opposite each other where the p.d. is exactly zero initially. The potentiometer C is therefore used to zero G.

3. After G is zeroed, support the magnet M vertically in a wooden clamp, and hold it above S as shown in Fig. 67a. Lower the magnet until it is close to S. The field $B$ due to M is now normal to the face of S.

(i) Vary $I$ in suitable steps, record the reading on A and the meter reading $\theta$ on G. Do not exceed the maximum permissible current. Record your results in a table, as below.

(ii) With a large current $I$, raise the magnet M so that the field $B$ at S becomes weaker. Observe the change in the deflection in G.

(iii) Reverse the magnet, so that the field $B$ at S is reversed. Observe the change in deflection in G.

(iv) Without altering the external circuit, replace the p-semiconductor by n-semiconductor. Observe the change in deflection in G. It will be necessary to re-zero G, with the magnet removed, before making an observation.

### MEASUREMENTS

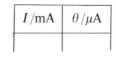

| $I$/mA | $\theta$/μA |
|--------|-------------|
|        |             |

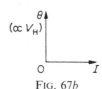

FIG. 67b

### GRAPH

Plot $\theta$ (proportional to $V_H$) against $I$.

### CONCLUSIONS

(a) From your graph, state how the Hall voltage $V_H$ is related to the current $I$ for a given field $B$.

(*b*) From your results in (ii) and (iii), say how $V_{\mathrm{H}}$ varies when $B$ is decreased for a given current and when the direction of $B$ is reversed.

(*c*) If the majority charge carriers in a n-semiconductor are electrons (negative charges), what conclusion do you draw from your result in (iv)?

## B. INVESTIGATION OF *B* INSIDE SOLENOID USING HALL PROBE

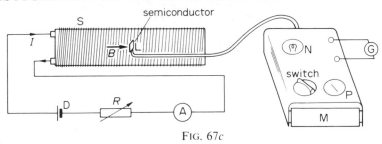

FIG. 67*c*

APPARATUS

Solenoid S (or tubular rheostat of a few ohms), Hall probe (e.g. UNILAB) and accessories, 12 V d.c. supply D, rheostat $R$ (10 Ω), meter A (0–10 A), Scalamp galvanometer G.

METHOD

1. Arrange the solenoid S in series with D, $R$ and A, as shown.

2. Place the Hall probe inside S, so that the bent end L, to which the semiconductor slice is attached, is in the *middle* of S. To obtain a Hall voltage across the slice, the solenoid axis must be perpendicular to L—$B$ is then normal to the semiconductor face.

3. The probe is connected to a box containing a battery M for supplying current to the semiconductor, an indicator lamp N, and a potentiometer P. Connect the Scalamp G as shown, and zero the reading (start with the *least* sensitive range of G), as explained on p. 160.

4. Switch on the current for the solenoid S. Adjust $R$ until a reasonably large deflection $\theta$ is obtained in G. Record the current $I$ in A, and $\theta$. Vary $I$ in suitable steps, note the new deflections $\theta$, and enter your results in a table, as below.

5. For a given high current $I$, observe how the deflection $\theta$ changes as the semiconductor L is moved towards one end of the solenoid S. How does the value of $B$ at the end compare with the value of $B$ in the middle of S?

MEASUREMENTS

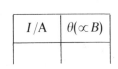

| $I$ /A | $\theta(\propto B)$ |
|---|---|
|  |  |

FIG. 67*d*

GRAPH

Plot $\theta$ (proportional to $B$) against $I$, from zero.

CONCLUSION

From your graph, say how $B$ varies with $I$ for a given number of turns per metre of the solenoid. Is it certain, from the results you have, that $\theta$ is proportional to $B$?

ADDITIONAL

(i) Using solenoids with different numbers $n$ of turns per metre connected in series, compare the field values $B$ in the middle of solenoids carrying the same current.

(ii) Investigate how $B$ varies from the middle of S to one end of the solenoid, for a given value of $I$.

# EXPERIMENT 68

# Determination of the Faraday Constant and Avogadro Constant

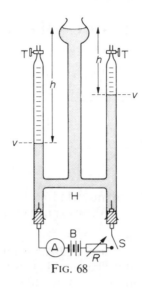

Fig. 68

## APPARATUS

Hoffmann voltameter, ammeter A, 6 V accumulator B, rheostat $R$, switch S, stop-watch, thermometer, hydrometer.

## METHOD

Fill the voltameter with sulphuric acid of relative density about 1·25. Connect the terminals attached to the platinum electrodes to the switch, accumulator, and rheostat, and adjust the rheostat to pass a current of about 1 A. Check that the gases are being evolved freely, and are not mixing *via* the horizontal connecting tube H. Switch off, allow all the bubbles formed to rise, and then allow the hydrogen and oxygen to escape by opening the taps T, until both columns are completely filled with electrolyte. Switch on again, starting the watch, and read the ammeter. When the level in the tube collecting hydrogen (connected to the negative pole of the battery) is near the horizontal tube H, switch off the current, and stop the watch. Note the time for which the current passed. Allow all the bubbles formed to rise, and then read the volumes ($v$) of each gas collected from the graduations on the collecting tubes. Measure the difference in height ($h$) between the level in each tube and that in the reservoir.

Measure the atmospheric pressure with a Fortin barometer, and the room temperature. Look up in tables the saturation vapour pressure of water at room temperature. Pour off some of the electrolyte, and measure its density with a hydrometer.

162

# MEASUREMENTS

Current $I =$ .. A  
Time $t$   = .. s

Density of electrolyte   = .. $\mathrm{kg\,m^{-3}}$  
Room temperature     = .. °C  
S.V.P. of water at .. °C= .. mmHg  
Barometric pressure $A$  = .. mmHg

| | Hydrogen | Oxygen |
|---|---|---|
| Volume of gas, $v/\mathrm{cm^3}$ | | |
| Head $h$ of electrolyte /mm | | |

From tables,  
Density of $H_2$ at s.t.p. $=0.0899\,\mathrm{kg\,m^{-3}}$  
„    „ $O_2$ „   „  $=1.429\,\mathrm{kg\,m^{-3}}$

# CALCULATION

The partial pressure $p$ of hydrogen$=\left[\, A+h\times\dfrac{\text{Density of electrolyte}}{\text{Density of mercury}}-\dfrac{\text{s.v.p. of water at}}{\text{room temperature}}\right]$

If the volume of hydrogen collected at $t$ in °C and pressure $p$ in mmHg is $v$ in $\mathrm{cm^3}$, then volume $V$ at s.t.p. is

$$V=v\times\frac{273}{273+t}\times\frac{p}{760}\ ..\ \mathrm{cm^3}=\ ..\ \times10^{-6}\,\mathrm{m^3}$$

Mass of hydrogen collected, $m=V\rho$, where $V$ is the volume at s.t.p. and $\rho$ is the density of hydrogen at s.t.p., $0.0899\,\mathrm{kg\,m^{-3}}$.

*Faraday Constant, F.* The mass $m$ in kg of hydrogen is deposited by a charge $Q=I\times t$, where $I$ is the current in ampere and $t$ is the time in second. So 2 moles of hydrogen atoms, mass $2\times10^{-3}$ kg, is deposited by a charge $Q$ given by

$$Q=\frac{2\times10^{-3}}{m}\times I.t=\ ..\ \mathrm{C}$$

$$\therefore\ F=\frac{Q}{2}=\frac{1}{2}\times\frac{2\times10^{-3}}{m}\times I.t=\frac{10^{-3}}{m}\times I.t=\ ..\ \mathrm{C\,mol^{-1}}$$

*Avogadro Constant, $N_A$.* Since $N_A e=F$, where $e$ is the electronic charge, $1.6\times10^{-19}\,\mathrm{C}$,

$$\therefore\ N_A=\frac{F}{e}=\frac{F}{1.6\times10^{-19}}=\ ..\ \mathrm{mol^{-1}}$$

# CONCLUSION

The Faraday constant, $F=$ .. $\mathrm{C\ mol^{-1}}$.  
The Avogadro constant, $N_A=$ .. $\mathrm{mol^{-1}}$.

# ERRORS. ORDER OF ACCURACY

1. Small errors in the heads $h$, the room temperature in °C, and the s.v.p. of water at room temperature will have little effect on the result, as they are added to (or subtracted from) much larger quantities.

2. Errors occur in the measurements of $v$ and $t$, and in reading the current $I$ and barometric pressure $A$. The percentage error in $F=$ sum of percentage errors in $v$, $I$ and $t$.

# EXPERIMENT 69

# Capacitance measurement using a Reed Switch

## APPARATUS

Reed switch unit, 6 V battery, 0–100 $\mu$A meter, 6·3 V a.c. supply with suitable rheostat (or audio-frequency oscillator), two capacitors each 0·1 $\mu$F, 2·2 k$\Omega$ resistor $R$.

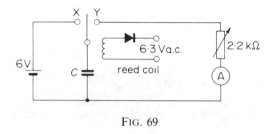

FIG. 69

## METHOD

The reed switch is a double switch which charges a capacitor $C$ through contact at X and then discharges $C$ through the meter A through contact at Y (Fig. 69). The rate of charge–discharge is determined by the frequency of the a.c. supply connected to the reed relay coil. A diode in the coil circuit prevents the reed vibrating at twice the frequency of the supply.

1. Connect the a.c. supply to the reed switch coil and adjust the rheostat so that the reed can be clearly heard vibrating. Connect the circuit shown in Fig. 69, using one 0·1 $\mu$F capacitor $C_1$ for $C$. Record the current $I$ flowing through the meter.

2. Replace $C_1$ by the other 0·1 $\mu$F capacitor, $C_2$. Record the current flowing with $C_2$.

3. (a) Replace the capacitor by $C_1$ and $C_2$ in series. Record the current flowing. (b) Now place $C_1$ and $C_2$ in parallel. Record the new current flowing.

## MEASUREMENTS

| Capacitor, $C$ | Current, $I/\mu$A |
|---|---|
| $C_1$ | |
| $C_2$ | |
| $C_1$ and $C_2$ – series | |
| $C_1$ and $C_2$ – parallel | |

## CALCULATIONS

In the circuit of Fig. 69, the capacitor $C$ is charged and discharged 50 times per second. So the total charge per second through the meter, or current $I$, is $50 \times CV$.

$$\therefore\ I = 50\ CV$$

$$\therefore\ C = \frac{I}{50V}$$

164

*Assuming C* is fully charged by the 6 V battery and fully discharged each time, calculate $C_1$, $C_2$ and their total capacitance in series and parallel from the measured values of $I$ in the table. ($C$ is in farad, F, if $V$ is in volt and $I$ in ampere.) Enter the results in the table.

## CONCLUSION

The capacitors of nominal value $0.1\,\mu F$ were found respectively to be .. $\mu F$, and .. $\mu F$. In series their combined capacitance was .. $\mu F$; in parallel their combined capacitance was .. $\mu F$.

## ADDITIONAL (optional)

A signal generator with low-impedance output can be used to drive the reed switch at frequencies up to 300 Hz. If the above equations are correct, altering the frequency by a factor of 2 should also alter the current $I$ by a factor of 2. Test this hypothesis, and so the validity of the equation for $C$.

## ORDER OF ACCURACY

Compare (i) the measured values of the two capacitors with their respective nominal $0.1\,\mu F$ values, and (ii) the measured values of the two in series and parallel with the values calculated from, respectively, $1/C = 1/C_1 + 1/C_2$ and $C = C_1 + C_2$. Comment on the factors in the experiment which affect the accuracy of the results.

## QUESTIONS

1. The $2.2\,k\Omega$ resistor is included in the circuit to 'protect' the meter. What would be the effect if the magnitude of the resistor was greatly increased?
2. What determines how well charged the capacitor $C$ becomes on connection to the battery at X?
3. If the frequency of operation of the reed switch could be changed, what would be the effect of (*a*) increasing the frequency, (*b*) decreasing the frequency?

# EXPERIMENT 70

## Measurement of Permittivity of Air

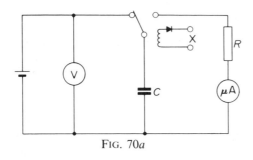

FIG. 70*a*

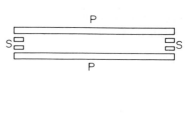

FIG. 70*b*

### APPARATUS

Flat capacitor plates P (e.g. Griffin & George, L 81–260); small polythene spacers S about 1 mm thick; 1 kg mass W; battery or other smooth d.c. supply 0–10 V or more; reed switch (e.g. Griffin & George, L 81–262); signal generator with low-impedance output; protective resistor R, 100 kΩ; sensitive light-beam galvanometer (e.g. Scalamp or Edspot); voltmeter 0–10 V; vernier calipers, metre rule; leads.

### METHOD

1. *General.* The circuit for the experiment is shown in Fig. 70*a*. A reed switch (see also Expt. 69) is used alternately to charge the capacitor C from the battery, to a voltage read on the voltmeter V, and discharge it via a protective resistor R through a sensitive microammeter. In this experiment the charge stored is small, so that a high frequency of charge and discharge is needed to give a detectable current in the meter. The reed switch can be driven from a signal generator, using the low-impedance output, up to about 400 Hz.

2. *Capacitance and spacing.* Use small polythene spacers (10 mm square, 1 mm thick) at the corners of the two plates, to keep the plates apart. The spacing can be varied in uniform steps, by stacking spacers on one another, as in Fig. 70*b*. The largest spacing used ought not to be greater than about $\frac{1}{20}$ of the dimensions of the plates, to avoid error due to edge effects.

Increase the frequency of the signal generator driving the reed switch, noting that the current in the meter increases. The current is given by $I = fQ$, where Q is the charge transferred at each switch operation, and $f$ is the frequency of vibration of the reed. At around 400 Hz the reed switch will cease to operate reliably, and the current will not increase with frequency. Use the highest frequency of reliable operation.

Record the current I in the meter for several values of the spacing, noting the number (n) of spacers stacked to produce this spacing. Record also the charging voltage V, and the frequency $f$ at which the switch is driven, keeping these constant. Measure and record the thickness t of a spacer S using vernier calipers, and the width w of the plates.

*Precaution.* While adding or taking away spacers, disconnect the battery, to avoid the chance that the battery is connected directly to the meter.

### MEASUREMENTS

Charging voltage $V$       = .. V  
Frequency of reed switch $f$ = .. Hz  
Thickness of a spacer $t$    = .. mm  
Width of plates $w$       = .. m

| Number of spacers $n$ | | |
|---|---|---|
| Current $I$ ($\mu$A) | | |

## GRAPH

Plot a graph of the current $I$ against $1/n$, where $n$ is the number of spacers used (Fig. 70c).

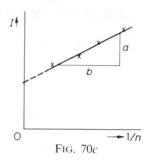

FIG. 70c

The current is proportional to the charge at a fixed voltage and hence to the capacitance. Since the capacitance is inversely proportional to the spacing, the graph of $I$ vs. $1/n$ may be expected to be a straight line. The line is unlikely to pass through the origin (very large spacing) because the plates have some fixed capacitance to the bench.

## CALCULATION

A value for $\varepsilon_0$ can be obtained from the slope $a/b$ of the graph. The capacitance $C$ is given by

$$C = \frac{\varepsilon_0 A}{d}$$

The plate area $A = w^2$ if the plates are square, and $d = nt$. Further, $I = fQ = fCV$, so that $C = I/fV$. Substituting in the above equation,

$$\frac{I}{fV} = \frac{\varepsilon_0 w^2}{nt}$$

So

$$I = \frac{\varepsilon_0 w^2 fV}{t}\left(\frac{1}{n}\right)$$

Hence the slope $a/b$ of the straight-line graph of $I$ v. $(1/n)$ is given by

$$\frac{a}{b} = \frac{\varepsilon_0 w^2 fV}{t}$$

$$\therefore\ \varepsilon_0 = \frac{t}{w^2 fV} \times \left(\frac{a}{b}\right)$$

Calculate $\varepsilon_0$ from this relation.

## CONCLUSION

The value of $\varepsilon_0$ obtained from this experiment was .. $F\,m^{-1}$.

## QUESTIONS

1. Many measured quantities enter the expression for $\varepsilon_0$. Which of them are known least accurately? Can you estimate the percentage accuracy with which they are known? It is unlikely that your value of $\varepsilon_0$ will be accurate to much better than 10%.

2. How would you investigate whether the charge on the capacitor was proportional to the charging voltage?

3. If the protective resistor $R$ is too big, the discharge will be slow and will not be complete before the reed switch changes over. The discharge takes a time of the order $RC$, while the time available for it is of the order $1/f$. Estimate $C$, calculate $RC$, and consider whether $RC$ is much smaller than $1/f$.

# EXPERIMENT 71

# Investigation of Parallel Plate Capacitance using Electrometer

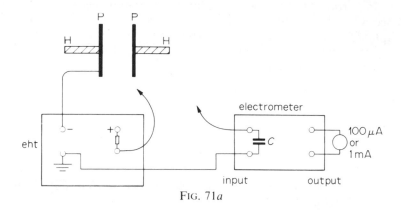

FIG. 71a

## APPARATUS

Metal plates with insulating handles (e.g. Griffin & George, N 11–700); E.H.T. power supply 0–5 kV with safety resistor; electrometer or d.c. amplifier (e.g. Griffin & George, L 91–400), with appropriate meter for indicating output (0–100 $\mu$A, or 0–1 mA); retort stands and clamps, input capacitor for electrometer (unless built in) 0·01 $\mu$F; leads; 1·5 V dry cell, potentiometer 5 k$\Omega$; voltmeter 0–1 V.

## METHOD

1. *General.* The electrometer will indicate voltages across its input in the range 0–1 V. Its input resistance is very high (perhaps $10^{13}\,\Omega$), so if a capacitor $C$ is connected across the input, any charge it is given leaks away very slowly. Hence there is an almost constant voltage across $C$, proportional to the charge put on it.

Thus the electrometer can be used to measure charge. We use it to investigate the charge stored on parallel plates.

2. *Calibration.* The electrometer consists of a small amplifier, which converts the voltage across its input to a current in a meter across the output. The electrometer can first be calibrated, or adjusted, so that a particular output current corresponds to a known input voltage. Use the circuit in Fig. 71b, in which the potentiometer $R$ can be a 5 k$\Omega$ radio potentiometer.

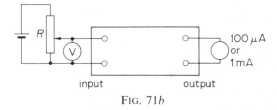

FIG. 71b

Adjust the potentiometer $R$ so that the voltmeter across the input reads 1 V. Alter the sensitivity control on the electrometer until the output current has a convenient value (100 $\mu$A or 1 mA). Reduce the voltage across the input to 0·5 V, and see if the output current is halved. (If there is evidence of non-linearity, it may be advisable to plot a calibration curve, but this should be unnecessary. Non-linearity could be due to the batteries driving the electrometer being run down, when they should be replaced.) Output currents can now be converted directly into input voltages.

168

3. *Investigations.* Support the plates P by their insulating handles H (see Fig. 71*a*) in retort stands, and arrange the plates parallel and about 10 mm apart.

Connect one plate to the negative side of the EHT supply, and to the earth terminal on the supply. Connect the earth terminal to the earthed side of the electrometer, as in Fig. 71*a*.

*Either* switch the electrometer to measure charge in the range $0$–$10^{-8}$ C, with an inbuilt capacitor of $0.01\ \mu$F across the input, *or* connect a polystyrene capacitor of this value across the input. Short the input for a moment, and check that the output stays at zero afterwards. If it does not, look for stray charges, which may be due to your own body, especially if you are wearing shoes with insulating soles, and working on or near a carpet of man-made fibres.

Now identify the positive terminal of the EHT supply which is in series with a $5\ \text{M}\Omega$ safety resistor. DO NOT connect to the unprotected terminal: the safety resistor will ensure that you do not obtain a shock from the supply.

(*a*) See if you can measure a charge put on the capacitor. Touch a lead from the protected positive terminal of the EHT supply, with the voltage set at about 500 V, onto the second plate. Remove this connection, and touch an insulated wire connected to the electrometer input, onto the plate (Fig. 71*a*). The meter connected to the electrometer should rise to a constant reading. Practise the operation until you can do it quickly. A stiff insulated wire inserted into the electrometer input, so that the wire can be made to touch the plate by tilting the electrometer, may be useful.

Wait for a minute after charging the plate before any measurements are made, and observe if the reading is now lower because charge has leaked away. If so, it may be advisable to dry the insulating handles with a hair dryer, after wiping away finger grease with a rag wetted with alcohol.

Now make a series of observations of the charge stored at a fixed spacing (say 10 mm), as the voltage is varied in the range 0–500 V. (If the power supply has no voltage indication, use a multimeter, but obtain advice about connecting it to the supply and avoiding shocks.)

(*b*) Then, at a fixed voltage, make a series of observations of the charge stored on the capacitor, at various spacings *d* in the range 10 mm to 50 mm.

MEASUREMENTS

An output current of .. $\mu$A (or mA) on the electrometer, corresponds to an input voltage of .. V.
Capacitance $C_\text{E}$ across electrometer input $= ..\ \mu$F.

(*a*)

| Charging voltage $V$/V | |
|---|---|
| Output reading | |
| Input voltage $V_\text{E}$/V | |
| Charge $Q$/C | |

(*b*)

| Spacing $d$/mm | |
|---|---|
| Output reading | |
| Input voltage $V_\text{E}$/V | |
| Charge $Q$/C | |

Convert each reading into the corresponding input voltage $V_\text{F}$, and calculate the charge $Q$ given to the electrometer, using $Q = C_\text{F} V_\text{F}$.

GRAPHS AND CONCLUSIONS

Plot graphs of (i) the charge $Q$ given to the electrometer, against the voltage $V$ across the capacitor, and of (ii) the charge $Q$ against $1/d$, where $d$ is the plate spacing.

The graph of $Q$ v. $V$ may be expected to be a straight line. Because there is about as large a stray capacitance to the bench as there is between the plates, the graph of $Q$ v. $1/d$ may be expected to be straight but may not pass through the origin. Draw what conclusions you can from your graphs.

# EXPERIMENT 72

## Charging and Discharging a Capacitor

### A. CHARGING A CAPACITOR

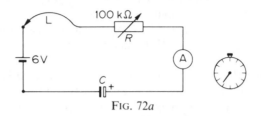

FIG. 72a

APPARATUS

6 V battery, 0–100 $\mu$A meter A, 100 k$\Omega$ rheostat $R$, 500 $\mu$F electrolytic capacitor $C$, stop-watch.

METHOD

*Preliminary.* Connect up the circuit in Fig. 72a, making sure that the positive terminal of the electrolytic capacitor is connected to the positive side of the 6 V supply. Remove the lead L after you have checked that a meter reading is obtained.

Now reconnect the lead L to the battery and observe the decreasing current. When the current is nearly zero, disconnect L and discharge the capacitor $C$ fully by short-circuiting it with a separate lead. Repeat the process, adjusting the rheostat so that the initial current, when L is connected to the battery, is about 100 $\mu$A.

*Experiment.* (1) With the capacitor $C$ fully discharged, reconnect the lead L to the battery and start the stop-watch when the current falls to 90 $\mu$A. Record the times from the start for the current to decrease to 80 $\mu$A, to 70 $\mu$A, and so on until the current reaches 30 $\mu$A. Repeat the experiment to obtain more reliable readings.

(2) Now increase the rheostat resistance so that the initial current when L is connected to the battery is about 75 $\mu$A. Repeat the above experiment, this time recording the times for the current to decrease from 70 $\mu$A down to 30 $\mu$A.

MEASUREMENTS

| Current /$\mu$A | Time /s |
|---|---|
| . . | . ., . .   Average = . . |

170

## GRAPH

Plot a graph of current, $I$, against time, $t$.

From your graphs, find the time taken for the current to fall:

    in (1), from 90 to $45\,\mu A = \ldots$ s; from 80 to $40\,\mu A = \ldots$ s,

    in (2), from 70 to $35\,\mu A = \ldots$ s; from 60 to $30\,\mu A = \ldots$ s

## CONCLUSIONS

In your conclusions discuss whether or not the current decay is exponential, and test this by looking at the times for the current to fall to half its value. Comment on the effect of the values of $CR$ on these times.

## QUESTIONS

1. What does the *area* beneath the graph and the time-axis represent?

2. How can you deduce from the graph whether $C$ charges more rapidly in experiment (1) than in experiment (2)?

3. When the lead L is first connected to the battery, does the value of the initial current depend on $C$ or $R$ or a combination of $C$ and $R$?

## NOTE

*Charging at constant rate.* By continuously altering the rheostat $R$, you can charge the capacitor $C$ in Fig. 71a by a constant current of say $80\,\mu A$. This requires practice. An oscilloscope is used across $C$ to measure the p.d. $V$. This is calibrated for d.c. voltage before connection so that a sensitivity of 1 V per 10 mm deflection of the spot or line trace is obtained (p. 194). The time $t$ for the voltage across $C$ to reach 1 V say (10 mm deflection from zero) is then measured. Since $Q = CV = I.t$, we have $C = I.t/V$. Thus knowing $I$, $t$ and $V$, then $C$ can be calculated. Compare the nominal value of $C$ with the calculated value.

(*Experiment continued overleaf*)

## 72. (Continued)

### B. DISCHARGING A CAPACITOR

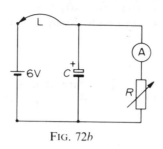

FIG. 72*b*

APPARATUS

6 V battery, 0–100 $\mu$A meter Z, 100 k$\Omega$ rheostat $R$, 500 $\mu$F electrolytic capacitor $C$, stop-watch.

METHOD

Connect up the circuit shown in Fig. 72*b*. By adjusting the rheostat $R$, arrange the initial current to be about 100 $\mu$A when the lead L makes contact with the battery terminal. Discharge $C$ completely with a separate lead. Now reconnect L briefly to the battery, and when the lead is disconnected so that $C$ discharges, start the stop-watch when the current reaches 90 $\mu$A. Record the time from the start for the current to fall to 80 $\mu$A, to 70 $\mu$A, and so on until 30 $\mu$A is reached.

Now increase the rheostat resistance so that the initial current is about 75 $\mu$A. Repeat the experiment, recording the times taken for the current to fall from 70, 60, ... to 30 $\mu$A.

MEASUREMENTS

| Current /$\mu$A | Time /s |
| --- | --- |
| | |

GRAPH

Plot a graph of current, $I$, against time, $t$, for your two sets of results. From your graph, determine the times for the current to fall from 90 to 45 $\mu$A and 80 to 40 $\mu$A in the first graph and from 70 to 35 $\mu$A and 60 to 30 $\mu$A in the second graph.

CONCLUSIONS

In your conclusions discuss whether or not the current decay is exponential, and test this by looking at the times for the current to fall to half its value. Comment on the effect of the values of $CR$ on these times.

# EXPERIMENT 73

# Calibration of Cathode-ray Oscilloscope as D.C. and A.C. Voltmeter

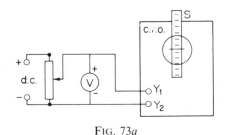

FIG. 73a

APPARATUS

Cathode-ray oscilloscope (see Note); well-smoothed d.c. supply; high-resistance potentiometer; high-resistance voltmeter, V, e.g. 0–10 V; transparent scale S, or perspex sheet with ruled markings; low-voltage transformer to supply a.c. at about 30 V (r.m.s.); a.c. voltmeter.

METHOD

1. *D.C. voltmeter.* Connect the h.t. supply across the potentiometer, so that a variable p.d. may be obtained, as in Fig. 73a. Connect the voltmeter V to one end of the potentiometer and to the sliding contact, observing the correct polarity, and connect the terminals of the voltmeter to the Y-plates of the cathode-ray oscillograph. (The terminals to the Y-plates may be marked $Y_1$ and $Y_2$, or Y and Earth E.) Switch off the d.c. supply.

Set the time-base control to a low frequency, and the brightness control to minimum, and switch on the cathode-ray oscilloscope. Increase the brightness until a trace appears and adjust the focus until the trace is sharp. Increase the time-base frequency if necessary, until the trace appears as a continuous line. *Do not allow a bright focused spot to remain stationary on the screen.*

Fix the scale S in a vertical position as close as possible to the face of the screen, and use the shift controls to set the trace on the screen on some convenient graduation near the centre of the scale.

Switch on the d.c. supply, and raise the p.d. applied to the Y-plates. Do not handle the leads to the supply when it is on. Record values of the voltage indicated by V and the corresponding displacements from its zero position of the trace on the screen. If a double-beam cathode-ray oscilloscope is being used the second beam may be left undeflected, and used as a 'zero' line.

Switch off the d.c. supply, reverse the connections to the Y-plates, and record a similar series of readings of V and displacement for deflections in the opposite direction.

2. *A.C. voltmeter.* Replace the d.c. supply by the a.c. supply from the transformer, and exchange the a.c. voltmeter for the d.c. meter used. Switch off the time base. Repeat the calibration measurements, this time recording the length *d* mm of the vertical line trace (twice the amplitude) and the corresponding reading *V* (r.m.s.) of the meter.

MEASUREMENTS

1. *D.C. voltmeter*

| Applied p.d. $V$ /V | | | | | |
|---|---|---|---|---|---|
| Displacement $y$/mm | | | | | |

## 2. A.C. voltmeter

| Applied p.d. $V$/V r.m.s. | | | | | |
|---|---|---|---|---|---|
| Length of trace $d$/mm | | | | | |

GRAPHS AND CALCULATION

Plot graphs of: (i) the displacement $y$ mm from zero of the trace v. $V$, the applied d.c. p.d.; (ii) the length $d$ mm of the trace v. $V$, the applied a.c. p.d. (Fig. 73$b$ (i), (ii)). Measure the slope $a/b$ of the straight part of each curve.

$$\text{d.c.: slope } (a/b) = \ .. \ \text{mm V}^{-1}$$
$$\text{a.c.: slope } (a/b) = \ .. \ \text{mm V}^{-1} \ \text{(r.m.s.)}$$

In each case note the range of values of $V$ for which the curve is a straight line.

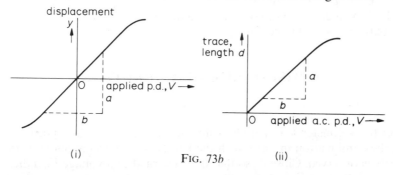

(i)                FIG. 73$b$              (ii)

CONCLUSION

The cathode-ray oscilloscope may be used as a voltmeter (i) for d.c. in the range .. V to .. V with sensitivity .. mm V$^{-1}$, (ii) for a.c. in the range .. V (r.m.s.) to .. V (r.m.s.) with sensitivity .. mm V$^{-1}$ (r.m.s.).

ERRORS

The accuracy with which the calibration may be performed is limited by the errors in reading the voltmeters, and by the accuracy with which a deflection or length of the trace may be measured. The error to be expected in using the cathode-ray oscilloscope as a voltmeter may be estimated by finding the smallest change in the applied p.d. which produces an observable change in the deflection or length of the trace. Parallax error may be expected to occur.

PROBLEMS

1. Discuss how: (a) the anode voltage applied to the cathode-ray oscilloscope; (b) the spacing of the deflection plates; and (c) the position of the plates affect the sensitivity.

2. If an internal amplifier of the cathode-ray oscilloscope was not used for a.c. the a.c. sensitivity is related to the d.c. sensitivity by a factor $2\sqrt{2}$ for a sine waveform, the trace length being twice the amplitude for a.c. Find whether your results are consistent with this relationship.

3. Discuss the advantages of the cathode-ray oscilloscope over moving-coil meters for observing transient or varying voltages. Attempt to measure the p.d. across the coil of an inductance when a current in it is switched off.

NOTE

If the oscilloscope has a switch for a.c. or d.c. inputs, switch it to d.c.

# EXPERIMENT 74

# Investigation of Discharge of Capacitor and Flashing Neon Circuit

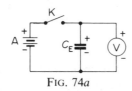

FIG. 74a

## APPARATUS

High-value electrolytic capacitor $C_E$, 500 $\mu$F–25 V working (see Note 1) and d.c. voltmeter 0–5 V of known resistance, e.g. 5000 $\Omega$ (see Note 2); 4 V accumulator A; key K; stop-watch; neon tube N (see Note 3); capacitor $C$, 0·1 $\mu$F; four known resistors, R, wireless type, 0·5 M$\Omega$, 5% or better; variable resistor $R_v$ 0–1 M$\Omega$ (e.g. linear 'volume control') cathode-ray oscillograph; well-smoothed d.c. supply 0–150 V or H.T. battery and 10 k$\Omega$ potentiometer P.

## 1. DISCHARGE OF CAPACITOR

### METHOD

Fig. 74a. Connect the voltmeter V in parallel with the capacitor $C_E$, their positive terminals being joined. Connect the combination in series with the key K to the accumulator A, making sure that the correct polarity is observed. Close K, so that $C_E$ is charged to a voltage $V_0$ indicated by the voltmeter, and record $V_0$. Release K and start the stop-watch at the same moment. As $C_E$ discharges through the resistance of the voltmeter, the reading falls, indicating the p.d. remaining across $C_E$. Choose some convenient lower reading $V$ and stop the watch as the voltmeter reaches $V$. Record the time $t$ taken, and repeat, having closed K again to charge $C_E$ to $V_0$ initially, for other values of the final reading $V$, covering as wide a range as possible. Record the value of $C_E$ marked on the capacitor and the resistance $R$ of the voltmeter.

### MEASUREMENTS

| P.d. across $C_E$ $V$ /V | $V_0=$ | | | | |
|---|---|---|---|---|---|
| Time $t$ /s | 0·0 | | | | |

Resistance of voltmeter $R=$ .. $\Omega$
$C_E=$ .. $\mu$F (nominal)

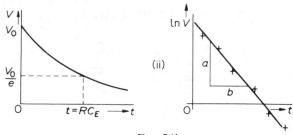

FIG. 74*b*

(i) Plot a graph of $V$ v. $t$ (Fig. 74*b* (i)). From $V = V_0 e^{-t/CR}$, where $V_0$ is the initial p.d. across the capacitor, $V = V_0 e^{-1} = 0.368 V_0$ after a time $CR$, known as the *time constant* of the circuit. From the graph,

$$V = 0.368 V_0 = \ .. \ V$$

after a time

$$t = \ .. \ \text{s.}$$

Estimate $C_E$ from

$$C_E = \frac{t}{R}$$

and compare with the value marked on the capacitor.

(ii) From $V = V_0 e^{-t/CR}$,

$$\ln V = \ln V_0 - \frac{t}{CR}$$

Plot $\ln V$ v. $t$ (Fig. 74*b* (ii)). Find whether a straight line can be drawn through the plotted points, thus testing the exponential law between $V$ and $t$. Measure the slope or gradient, $-a/b$, of the line:

$$\text{slope}\left(-\frac{a}{b}\right) = \ .. \ = -\frac{1}{CR}$$

$$\therefore \ C = C_E = R \times \frac{a}{b} = \ .. \ F$$

Discuss which estimate of $C_E$ is likely to be the more accurate, and the effect of damping and inertia of the meter movement.

CONCLUSIONS

Time constant $CR = \ .. \ \text{s.} \ C_E = \ .. \ \mu\text{F.}$ The graph of $\ln V$ v. $t$ is consistent with the relation $..$ for the discharge of a capacitor.

(*Experiment continued overleaf*)

## 74. (*Continued*)

### 2. NEON TUBE CIRCUIT

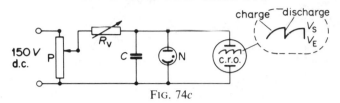

FIG. 74c

METHOD

Fig. 74c. Connect the neon tube N and capacitor $C$ in parallel, and join the leads from N to the Y-input (or Y and E) of the cathode-ray oscilloscope. Connect one side of $C$ in series with the variable resistance $R_v$, set at maximum resistance, and the sliding contact of the potentiometer P across the h.t. supply. Join the other side of $C$ to one end of P. Raise the voltage applied to N until it flashes regularly. Adjust the time base of the cathode-ray oscilloscope and reduce $R_v$ until a steady pattern is obtained as indicated in Fig. 74c. Reverse the Y-connections if necessary. Explain the form of the pattern, remembering that the neon gas in the tube will glow—offering low resistance—if the voltage across it has once risen to the striking potential and is still above the (lower) extinction potential at which the glow ceases.

Vary $R_v$ and also the applied potential and observe the changes in the trace. Explain the changes in the flashing rate and in the steepness of the rising part of the trace. Verify that the vertical distance between the 'peak' and 'trough' of the trace, which is proportional to the difference between the striking and extinction potentials, remains constant.

*Resistance comparison.* Set $R_v$ at maximum and adjust the voltage until flashes occur at about 2-sec. intervals. Record the time for, say, 50 flashes. Repeat with first one, then two, three and four of the known resistors $R$ in series in place of $R_v$. Note the value marked on $C$.

### MEASUREMENTS

| $C = ..\ \mu F$ | $R_v$ at max. | $R = ..\ \Omega$ | $2 \times R = ..\ \Omega$ | $3 \times R = ..\ \Omega$ | $4 \times R = ..\ \Omega$ |
|---|---|---|---|---|---|
| Time /s | | | | | |
| No. of flashes | | | | | |
| Time $t$ for one flash /s | $t_v =$ | | | | |
| $(R/t)/\Omega\,s^{-1}$ | | | | | |

CALCULATION

1. Calculate the value of $R/t$ for each of the combinations of known resistors in series, and examine whether $R/t$ is constant within experimental error. Explain why the value of the time-constant $CR$ differs from $t$.

2. Find the value of $R_v$ (at maximum) from the time $t_v$ between flashes with $R_v$ connected to $C$: $R_v = t_v \times$ average value of $R/t = ..\ \Omega$.

178

CONCLUSIONS

1. For fixed $C$ and applied voltage, the time between flashes is .. to the resistance in series with $C$.

2. $R_v = ..$ $\Omega$, set at its maximum value.

PROBLEM

Repeat one measurement of $t$, with a d.c. voltmeter, 0–150 V, connected across the potentiometer. Record the applied voltage $V$. Remove $C$, place the voltmeter across N, and connect 100 k$\Omega$ in series with the power supply. Raise the voltage until N strikes, and record $V_S$, the striking potential. Decrease $V$ and record $V_E$, the extinction potential. Record the applied voltage $V$ used above, and verify by substitution that

$$t = CR \ln \frac{V - V_E}{V - V_S}$$

using the above results.

NOTES

1. High capacitors, such as that described, are available at low cost from radio component dealers.

2. A 0–1 mA meter and appropriate series resistance will serve in place of the voltmeter.

3. Neon tubes intended to be connected direct to h.t. have a ballast resistor sealed into the cap, and this must be removed. Alternatively, small neon bulbs, without any resistor sealed in, are available from radio dealers.

# EXPERIMENT 75

# Reactance of Capacitor.  Series $C$–$R$ A.C. Circuit

APPARATUS

Cathode-ray oscilloscope, audio-frequency oscillator with low-impedance output, non-inductive resistance $R$ such as $330\,\Omega$ and capacitance $C$ of $1\,\mu\mathrm{F}$.

*(The oscilloscope and oscillator must not have a common earth.)*

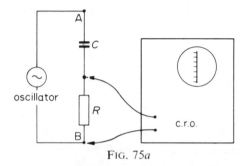

FIG. 75a

METHOD

1. Connect the circuit shown and adjust the frequency of the oscillator to $1000\,\mathrm{Hz}$ (Fig. 75a). Join
the oscilloscope across the resistor $R$, and adjust the gain until a vertical line is obtained *without* the
time base, which has a suitable length $l_1$ such as $50\,\mathrm{mm}$.

Keeping the gain unaltered, connect the oscilloscope across (a) $C$ and record the length $l_2$ of the
vertical line, and then (b) $C$ and $R$ in series, terminals A and B, and record the length $l_3$ of the vertical
line.

2. If you are using a double-beam oscilloscope, you can observe the waveforms of the voltages $V_R$
and $V_C$ and any phase difference between them.

3. Switch off the time base. Repeat the measurements of $l_1$ ($R$), $l_2$ ($C$) and $l_3$ ($R$ and $C$ in series) over
a range of frequencies from 1000 to $100\,\mathrm{Hz}$. At some frequency the length $l_2$ for $C$ will be greater
than the length $l_1$ for $R$. For this frequency, readjust the gain so that the length $l_2$ is say $50\,\mathrm{mm}$ and
with the gain constant, record $l_1$ for $R$ and $l_3$ for $C$ and $R$ in series, terminals A and B. Repeat for
lower frequencies, keeping the gain constant when $l_1$, $l_2$ and $l_3$ are measured.

MEASUREMENTS

| Frequency $f$/Hz | $l_1$/mm $(V_R)$ | $l_2$/mm $(V_C)$ | $l_3$/mm $(V)$ | $l_1/l_2$ | $\sqrt{l_1{}^2+l_2{}^2}$ |
|---|---|---|---|---|---|
| | | | | | |

GRAPH AND CALCULATION

(1) *Capacitance calculation.* Calculate the ratio $l_1/l_2$ and enter the values at different frequencies in the Table.

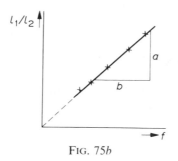

FIG. 75*b*

Plot the graph of $l_1/l_2$ against $f$ (Fig. 75*b*). Draw the best straight line through the points and measure the gradient $g(=a/b)$ of the line.

Since $V_C/V_R = X_C/R = 1/2\pi f CR = l_2/l_1$,

$$\therefore\ C = \left(\frac{l_1}{l_2}/f\right) \times \frac{1}{2\pi R} = \frac{g}{2\pi R} \qquad \cdots \cdots \cdots \cdots \text{(i)}$$

(*a*) From (i), calculate $C$ using the gradient $g$ and $R$.

(*b*) From your graph, obtain the frequency $f$ where $l_1/l_2 = 1$. Then calculate $C$, from (i), using the relation $C = 1/2\pi Rf$. Compare your value with that in (*a*).

(2) *Voltage relation.* Calculate the values of $\sqrt{l_1{}^2 + l_2{}^2}$ at each frequency and enter the results in the Table. Compare the result at each frequency with the corresponding value of $l_3$.

Construct a triangle of sides $l_1(V_R)$, $l_2(V_C)$ and $l_3(V)$ for one or more frequencies. What is the phase angle between $V_C$ and $V_R$?

CONCLUSION

The capacitance $C = \ ..\ \mu F$.

If $V_C$ and $V_R$ are the respective voltages across $C$ and $R$ in series, the voltage $V$ across both components $= \ ..$

The phase difference between $V_C$ and $V_R = \ ..°$

NOTE

In practice, a capacitor has a loss of energy in an a.c. circuit which may be considered equivalent to a resistor $r$ in parallel with $C$. Does your graph of $l_1/l_2$ against $f$ suggest a high or a low value of $r$ for your capacitor?

# EXPERIMENT 76

# Reactance of Inductor.  Series $L–R$ A.C. Circuit

## APPARATUS

Cathode-ray oscilloscope, audio-frequency oscillator with low-impedance output, non-inductive known resistance $R$ such as $330\,\Omega$ and inductance $L$ of about $100\,mH$ ($0.1\,H$). Coils from a demonstration transformer (Griffin & George, L 83–164/078) or inductance coils (Griffin & George, L 88–560) or solenoids (Griffin & George, L 83–750) are suitable.

   (*The oscilloscope and oscillator must not have a common earth.*)

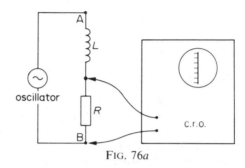

FIG. 76*a*

## METHOD

1. Connect up the circuit shown in Fig. 76*a*; use an oscillator frequency of $100\,Hz$. Connect the oscilloscope across $R$ and adjust the gain to give a suitably long vertical line of say $50\,mm$ *without* the time base. Record the length $l_1$ of the line.

   Keeping the gain constant, connect the oscilloscope (*a*) across the inductor $L$ and measure the length $l_2$ of the vertical line, and then (*b*) across $L$ and $R$ in series, terminals A and B, and measure the length $l_3$ of the vertical line. Record $l_2$ and $l_3$.

2. If you are using a double-beam oscilloscope, you can observe the waveforms of the voltages $V_R$ and $V_L$ and any phase difference between $V_R$ and $V_L$.

3. Switch off the time base. Repeat the measurement of $l_1$ ($R$), $l_2$ ($L$) and $l_3$ (series $L$, $R$) over a range of frequencies from $100\,Hz$ to $1000\,Hz$ in steps of say $200\,Hz$. At some frequency the length $l_2$ ($L$) will become greater than the length $l_1$ ($R$). For this frequency, readjust the gain so that $l_2$, the *inductor* length, is a suitably long line such as $50\,mm$, and then measure $l_1$ ($R$) and $l_3$ (series $L$, $R$) keeping the gain constant for all three measurements.

   At higher frequencies, alter the gain so that $l_2$ ($L$) is say $50\,mm$ long, and keeping the gain constant, measure $l_1$ ($R$) and $l_3$ (series $L$, $R$).

## MEASUREMENTS

| Frequency $f$/Hz | $l_1$/mm ($V_R$) | $l_2$/mm ($V_L$) | $l_3$/mm ($V$) | $l_2/l_1$ | $\sqrt{(l_1{}^2 + l_2{}^2)}$ |
|---|---|---|---|---|---|
|  |  |  |  |  |  |

## GRAPH AND CALCULATION

(1) *Inductance calculation.* Calculate the ratio $l_2/l_1$ and enter the values at the different frequencies $f$ in the Table. Plot $l_2/l_1$ against $f$ (Fig. 76b). Draw the best straight line which passes through the origin and the points at *higher* frequencies, as shown. Measure the gradient $g (=a/b)$ of the line.

Since $V_L/V_R = X_L/R = 2\pi f L/R = l_2/l_1$,

$$L = \frac{(l_2/l_1)}{f} \times \frac{R}{2\pi} \qquad \cdots \cdots \cdots \cdots \cdots \quad \text{(i)}$$

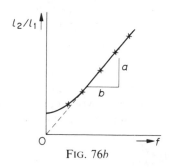

FIG. 76b

(a) From (i), $L = g \times R/2\pi$. Calculate $L$ using the gradient value $g$ and $R$.

(b) From (i), when $l_2/l_1 = 1$, then $L = R/2\pi f$. From the graph, read off the value of $f$ where $l_2/l_1 = 1$, and calculate $L$ from $L = R/2\pi f$. Discuss briefly any discrepancy with the value of $L$ from (a).

(2) *Voltage relation.* Calculate the values of $\sqrt{l_1^2 + l_2^2}$ at each frequency and enter the results in the Table. Compare the result with the value of $l_3$ for each frequency.

Construct a triangle of sides $l_1(V_R)$, $l_2(V_L)$ and $l_3(V)$ for a low and a high frequency. What is the phase angle between $V_L$ and $V_R$ in each case? Comment.

## CONCLUSION

The inductance $L = \ldots$ H.

If $V_R$ and $V_L$ are the voltages across $R$ and $L$ respectively, then the voltage $V$ across both $R$ and $L$ in series is given by $V = \ldots$

The phase difference between $V_R$ and $V_L$ at a high frequency is $\ldots^\circ$.

## NOTE

The inductor coil has some d.c. resistance, $r$ say. Find the value of $r$ by a voltmeter–ammeter method, using a low-voltage d.c. supply and rheostat. What effect has this value $r$ on (i) the graph in Fig. 76b, (ii) the value of $V$, the total voltage across the inductor and resistor in series?

183

# EXPERIMENT 77

# Investigation of Series Resonance Circuit

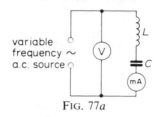

variable
frequency ~
a.c. source

FIG. 77a

## APPARATUS

Coil $L$, about $0.1\,H$ (as Expt. 76); paper or mica capacitor $C$, $0.1\,\mu F$; cathode-ray oscilloscope; a.c. milliammeter mA, 0–10 mA; high-resistance a.c. voltmeter V, 0–10 V; non-inductive resistance box $R$, 0–500 $\Omega$; oscillator of low output impedance (or oscillator and amplifier); means of measuring d.c. resistance of coil.

## 1. CURRENT RESONANCE

### METHOD

Connect the coil $L$, capacitor $C$ and milliammeter mA in series and join them to the oscillator output (Fig. 77a). Connect the voltmeter V across the output. (The cathode-ray oscilloscope may be used in place of $V$—see Notes, p. 187.)

Vary the frequency over the range 300–3000 Hz and find the frequency at which the current is greatest. Adjust the output until mA reads near full scale (see Notes).

Record pairs of values of the current $I$ in mA and voltage $V$ in V at frequencies covering the above range. If time allows, repeat for another value of $C$, say $0.5\,\mu F$. Measure the d.c. resistance $r$ of the coil (e.g. by voltmeter–ammeter method or by metre bridge method, p. 138).

### MEASUREMENTS

| Output voltage, $V$/V r.m.s. | | | | |
|---|---|---|---|---|
| Current, $I$/A r.m.s. | | | | |
| Frequency, $f$/Hz | | | | |

$C = \ .. \ \mu F.$

Coil data for $L$ (if required): Total turns $N = \ ..$, mean radius $a = \ .. \ m$, length, $l = \ .. \ m$

d.c. resistance $r = \ .. \ \Omega.$

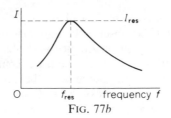

FIG. 77b

Plot a graph, as in Fig. 77b, of the current $I$ v. frequency $f$. Record from the graph the values of the maximum current $I_{res}$ and the frequency $f_{res}$ at which this occurs. Deduce the value of $Z_{res} = V/I_{res}$, where $V$ is the voltage when $I = I_{res}$.

$$f_{res} = \,.. \text{ Hz} \qquad Z_{res} = \,.. \text{ } \Omega$$

Theory shows that the resonant frequency

$$f_{res} = \frac{1}{2\pi\sqrt{LC}} \quad \text{and} \quad Z_{res} = r$$

Test these predictions by substituting the value of $C$, that of $L$ calculated from $\mu_0 N^2 A/l$ henry (where $\mu_0 = 4\pi \times 10^{-7} \text{ H m}^{-1}$, $A = \pi a^2$ in metre$^2$, $l = $ length in metre) and the measured value of $r$. Discuss any discrepancy, noting that the formula for $L$ holds only for a long narrow coil.

## 2. VECTOR DIAGRAM

METHOD

Connect $L$, $C$ and the resistance box $R$ set at $200\,\Omega$ to the output of the oscillator (Fig. 77c).

1. *Phase difference between $V_L$ and $V_C$.* Join the junction of $L$ and $C$ to the earth E terminal of the cathode-ray oscilloscope (or to both $Y_2$ and $X_2$ if the inputs are independent). Connect the other terminals of $L$ and $C$ to the Y ($Y_1$) and X ($X_1$) terminals of the cathode-ray oscilloscope. Switch off the time base.

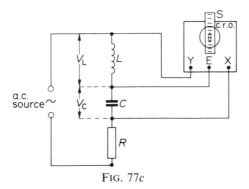

FIG. 77c

Set the oscillator frequency at a low value, when the voltage $V_L$ across $L$ will be small (see p. 183) and the voltage $V_C$ across $C$ will be large (p. 181). Check that the figure observed on the screen is consistent with these expectations. Increase the frequency, and observe the changes in the Lissajous' figure traced on the cathode-ray oscilloscope. Note that at resonance both $V_L$ and $V_C$ are large, and

(*Experiment continued overleaf*)

# 77. (*Continued*)

that the shape of the figure indicates that there is a phase angle of 180° between them. (If in doubt, replace $L$ and $C$ by resistors of $1000\,\Omega$ and observe the figure due to the voltages across these which are *in phase*.) Increase the frequency beyond resonance and observe the effect on the trace of the diminishing size of $V_C$ compared with that of $V_L$. Sketch the shape of the figure at each stage.

2. *Vector addition of $V_L$ and $V_C$.* Disconnect the lead to the cathode-ray oscilloscope X-input and connect the leads from Y and E ($Y_1$ and $Y_2$) across the $L$–$C$ series combination. Vary the frequency through resonance and observe that the voltage $V_{LC}$ across $L$ and $C$ reaches a minimum at resonance. Record the length $d_{LC}$ of the trace at the resonant frequency. Now connect the Y-input leads first across $L$ and then across $C$, and record the respective trace lengths $d_L$ and $d_C$ at the same frequency.

## MEASUREMENTS

### At Resonance

Trace length (proportional to voltage $V_{LC}$), $d_{LC} = \ldots$ mm
,,  ,,  (  ,,  ,,  ,,  $V_L$), $d_L$  $= \ldots$ mm
,,  ,,  (  ,,  ,,  ,,  $V_C$), $d_C$  $= \ldots$ mm

## VECTOR DIAGRAM

Fig. 77*d*. Draw a horizontal vector YD proportional to $d_{LC}$ and draw a vector YB of length proportional to $d_C$ perpendicular to YD. Find the direction YA of a vector proportional to $d_L$ such that its resultant with YB is YD. Find if the angle AYB is close to 180°. Explain how, because of their phase difference, $V_L$ and $V_C$ add vectorially to a *small* resultant, though $V_L$ and $V_C$ are each large.

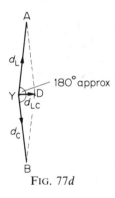

FIG. 77*d*

## CONCLUSIONS

1. *Resonant frequency.*   Measured value   $= \ldots$ Hz
    Calculated value   $= \ldots$ Hz
2. *Impedance at resonance.* Measured impedance$= \ldots\,\Omega$
    d.c. resistance of coil$= \ldots\,\Omega$
    Comments ...

3. Evidence that at resonance the voltages across $L$ and $C$ are large but differ in phase by nearly 180° is provided by ...

## PROBLEM

Measure the trace length $d_R$ due to the voltage across $R$ at resonance, and deduce:

$$X_L = \frac{d_L}{d_R}. R = \; .. \; \Omega; \quad X_C = \frac{d_C}{d_R}. R = \; .. \; \Omega; \quad r = \frac{d_{LC}}{d_R}. R = \; .. \; \Omega$$

Compare with the expected values

$$X_L = 2\pi f L; \quad X_C = \frac{1}{2\pi f C}; \quad r = \text{d.c. resistance of coil}$$

## NOTES

1. The cathode-ray oscillograph may be calibrated as an a.c. voltmeter as Expt. 73, p. 174, and may then be used in part 1. Calibration also avoids the assumption made in part 2 that the trace length is proportional to the voltage applied to the cathode-ray oscilloscope.

2. At resonance, the p.d. across $L$ or $C = I\sqrt{L/C} = 10^3 I$ for the components suggested. To avoid dangerously high voltages, $I$ must not exceed, say 50 mA. The meter specified in part 1, and the value of $R$ in part 2, are chosen with this in mind.

# PART SEVEN

# Atomic Physics including Electronics

# EXPERIMENT 78

# Junction Diode: Characteristic and Rectification

## APPARATUS

Diode OA81 or equivalent. 1·5 V battery, 500 Ω potentiometer, 0–1·5 V meter, 0–10 mA meter (forward bias), Scalamp galvanometer (reverse bias).

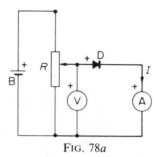

FIG. 78a

## METHOD

1. Connect up the circuit shown in Fig. 78a (forward bias). B is the battery, R is the potentiometer, D is the diode in forward bias, A is the current meter and V is the voltmeter. (*Care!* Read the manufacturer's catalogue for the maximum voltage and current in forward bias, so as not to damage the diode.)

2. Set the potentiometer R so that the voltage in V and the current in A are zero. Adjust R so that the voltage $V$ increases in suitable small steps such as 0·2 V from zero to the maximum such as 1 V, and record the values of $V$ and $I$ from the meters.

3. *Reverse* the diode D in the circuit and substitute the Scalamp in place of the meter A. Record the value of $I$ at a reverse voltage of 1 V.

## MEASUREMENTS

Forward bias

| voltage $V$/V | current $I$/mA |
|---|---|
| | |

Reverse bias      At 1 V, current $I = $ .. $\mu$A

## GRAPH

Plot a graph of $I$ against $V$ for forward bias. On the same axes, using negative values for $V$ and $I$, indicate roughly the small current for the reverse bias of 1 V and the form of the graph for reverse bias.

## CALCULATION

Using the graph of forward bias (Fig. 78b), calculate (i) the reciprocal, $\Delta V/\Delta I$, of the gradient of the graph at $V = 0.7$ V (this gives the 'a.c. resistance' of the diode at this voltage) and (ii) the ratio $V/I$ at $V = 0.7$ V, the d.c. resistance at this voltage.

a.c. resistance at $0.7$ V $= \, .. \, \Omega$

d.c. resistance at $0.7$ V $= \, .. \, \Omega$

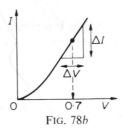

FIG. 78b

## CONCLUSION

Discuss the resistance of the junction diode in forward and reverse bias and whether the diode is an 'ohmic' or 'non-ohmic' component.

### Rectification

#### APPARATUS

Low-voltage a.c. supply $V$ such as $6.3$ V at $50$ Hz from mains *or* a.f. oscillator, diode D and series resistor $R$ of $33$ k$\Omega$, and capacitors $C = 0.1$, 1 and $4 \, \mu$F, oscilloscope (Fig. 78<i>c</i>).

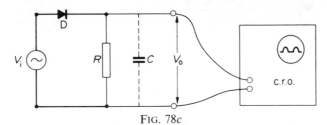

FIG. 78c

#### METHOD

Connect the *input* voltage $V_i$ to the oscilloscope and observe the waveform. Now connect the *output* voltage $V_o$ to the oscilloscope (a) across $R$, and then (b) across $R$ and $C$—vary $C$ from $0.1 \, \mu$F to $4 \, \mu$F and observe the effect on smoothing the output voltage. Fig. 78<i>c</i>. Is the smoothing better with the larger capacitors? If so, why?

# EXPERIMENT 79

## Investigation of Diode Valve Characteristics

### APPARATUS

Diode valve, suitable variable d.c. power supply P for anode voltages, heater supply (e.g. 6·3 V) $E_1$, milliammeter (mA), ammeter M, voltmeter V, rheostat $R$ (e.g. 10 Ω).

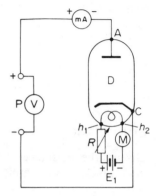

### METHOD

1. Connect the heater terminals $h_1$, $h_2$ to the rheostat $R$, ammeter M and heater supply $E_1$ (Fig. 79a). Connect the positive terminal of the power supply P to the anode A through the milliammeter mA and its negative terminal to the cathode C if the valve is indirectly heated, or, if directly heated, to the terminal $h_2$ joined to the negative terminal of $E_1$. Connect the voltmeter V across P to measure the anode voltage.

2. Set the rheostat $R$ so that the heater is operating at its normal voltage. By means of P, increase the anode voltage in steps, reading the milliammeter at each setting. Take care not to let the anode current rise above the maximum permissible value for the valve, or the anode will be overheated.

3. Reduce the heater current $(I_H)$ by means of $R$, and again take a series of anode voltage $(V_a)$ and current $(I_a)$ readings; take one set of readings with the heater only just glowing. Record the heater current each time.

### MEASUREMENTS

| | | | |
|---|---|---|---|
| $I_H = ..$ A | $V_a/V$ | |
| | $I_a/mA$ | |
| $I_H = ..$ A | $V_a/V$ | |
| | $I_a/mA$ | |
| $I_H = ..$ A | $V_a/V$ | |
| | $I_a/mA$ | |
| $I_H = ..$ A | $V_a/V$ | |
| | $I_a/mA$ | |

## GRAPHS

1. Plot $I_a$ against $V_a$ for the various heater currents $I_H$ (Fig. 79b). From the slope of the linear part of the graph $(XY)$,

$$\text{anode resistance, } R_a = \frac{dV_a}{dI_a} = \ .. \ \Omega$$

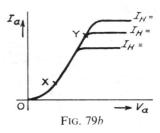

FIG. 79b

2. From the table of measurements, calculate $\log I_a$ and $\log V_a$ for the series of values at maximum heater current corresponding to the part OXY of the $I_a - V_a$ graph. Plot $\log I_a$ v. $\log V_a$, draw the best straight line through the points, and measure the slope or gradient of the line. Since $I_a = kV_a{}^n$ for this part of the graph, where $k$ is a constant, $\log I_a = \log k + n \log V_a$, and thus the gradient of the line is $n$.

## CONCLUSIONS

1. The anode resistance $R_a = \ .. \ \Omega$.
2. The relation between $I_a$ and $V_a$ has the form $I_a = kV_a{}^n$, where $n = \ ...$

## NOTES

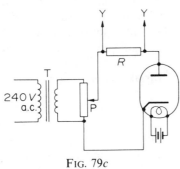

FIG. 79c

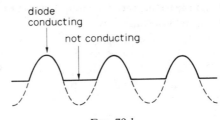

FIG. 79d

1. *Cathode-ray tube display of waveform of rectified current:*

The diode produces a rectified current of the order of a few milliamps. To convert this into a *voltage* large enough to deflect the beam of a cathode-ray tube, a resistance $R$ of about $10 \text{ k}\Omega$ is connected in series with the valve (Fig. 79c) and the voltage appearing across it is connected to the Y plates of the tube. A suitable alternating supply is obtained from a 24 V transformer T connected to a potentiometer P, as shown.

The time base is adjusted to give a steady pattern. Since the valve conducts only during the positive half-cycle of the alternating supply, the pattern will be as in Fig. 79d. Connect capacitors ranging from 0·1 to 10·0 $\mu$F in parallel with the resistance $R$ and observe the effect on the pattern. Explain this, remembering that the capacitor will discharge through $R$ when the valve is not conducting.

2. Many indirectly-heated diodes become overloaded before saturation. The GRD7 valve is suitable for this experiment.

193

# EXPERIMENT 80

## Investigation of Transistor Characteristics—Common Emitter Circuit

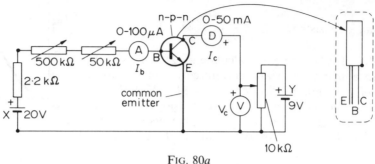

FIG. 80a

### APPARATUS

Transistor $n$–$p$–$n$ (e.g. BC 108) or similar; variable resistances (potentiometers) of 500, 50 and 10 kΩ respectively; 2·2 kΩ resistor; batteries providing about 20 V and 9 V respectively; high-resistance voltmeter V; microammeter A 0–100 $\mu$A, milliammeter D 0–50 mA.

### METHOD

1. Before starting, make sure your transistor is a $n$–$p$–$n$ type, i.e. the emitter E is $n$-type, the base B is $p$-type and the collector C is $n$-type semiconductor (see *Note* below if you have a $p$–$n$–$p$ transistor). Starting with the larger battery X, connect its NEGATIVE pole to the emitter terminal E (Fig. 80a). Join its positive pole to the base terminal B, with the 2·2, 500 and 50 kΩ variable resistors, and the microammeter A all in series, as shown.

Connect the smaller battery Y across the whole of the 10 kΩ variable resistor. Join the voltmeter V across the slider and one fixed terminal of the resistor, so as to measure a variable p.d. Connect the POSITIVE side of the variable p.d. to the collector terminal C through the milliammeter D, and the negative side of the variable p.d. to the emitter terminal E. Before starting, check again that the positive side of the battery X is joined to B, that the positive side of the variable p.d. is joined to C and that the emitter E is connected to the negative terminals of the batteries X and Y.

(*Note.* If a $p$–$n$–$p$ type transistor is used the battery connections shown in Fig. 80a must be *reversed*, i.e. the base B is negative in potential relative to E, and C is negative relative to E.)

2. *Output characteristic.* ($I_c$–$V_c$, $I_b$ constant.) Keeping the base current, $I_b$, constant at 40 $\mu$A on A, vary the potential $V_c$ of the collector in suitable steps from zero to about 8 V. Take a few readings in the 0–1 V range. Observe the collector current $I_c$ in D and the corresponding voltage $V_c$ on each occasion. If necessary, adjust the 500 kΩ (coarse adjustment) and 50 kΩ (fine adjustment) to keep the base current constant. Repeat with constant base currents of 60 and 80 $\mu$A respectively.

3. *Transfer characteristic.* ($I_c$–$I_b$, $V_c$ constant.) By means of the potential divider, adjust the collector voltage $V_c$ to about 4·5 V. With $V_c$ constant, increase the base current $I_b$ in suitable steps by altering the 500 or 50 kΩ resistors, and observe the corresponding collector current, $I_c$, until it reaches about 40 or 50 mA or other suitable value.

### MEASUREMENTS

$I_b = \; .. \; \mu$A

|  | $V_c$/V | | | | | | |
|---|---|---|---|---|---|---|---|
| Output characteristics: | $I_c$/mA | | | | | | |

194

$$V_c = \ .. \ \text{V}$$

| Transfer characteristic: | $I_b/\mu\text{A}$ | | | | | | | |
|---|---|---|---|---|---|---|---|---|
| | $I_c/\text{mA}$ | | | | | | | |

## GRAPHS

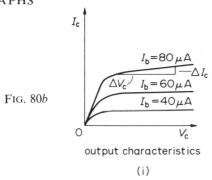

FIG. 80b

output characteristics

(i)

transfer characteristic

(ii)

Plot (i) $I_c$ v. $V_c$ ($I_b$ constant) (Fig. 80b(i))
(ii) $I_c$ v. $I_b$ ($V_c$ constant) (Fig. 80b(ii))

## CALCULATIONS

1. From the slope of the straight-line part of the graph beyond the knee of the graph, calculate the output impedance, $R_o$, for two different values of $I_b$ from $R_o = \Delta V_c/\Delta I_c$ ($I_b$ constant) (Fig. 80b(i)).

$$\text{Output impedance} = \ .., \ .. \ \Omega$$

2. From the $I_c$–$I_b$ graph, calculate the *current gain*, $\Delta I_c/\Delta I_b$, for $I_c = 30\,\text{mA}$ from the gradient of the line here (Fig. 80b(ii)).

$$\text{Current gain} = \ .. \ = \ .., \text{at } I_c = \ .. \ \text{mA}$$

## CONCLUSIONS

The output impedance of the common emitter circuit $= \ .. \ \Omega.$
The current gain of the common emitter circuit $\quad = \ .. \text{ at } I_c = \ .. \ \text{mA}.$

## PROBLEMS

1. From the $I_c$–$I_b$ graph, find the variation of current gain with collector current $I_c$, for a range of collector current from about 5 to 40 mA.
Plot the current gain v. $I_c$, and comment on your result.

2. In the *base circuit* of Fig. 80a remove the microammeter A and substitute an audio oscillator with a pair of earphones in series. Arrange for a note of frequency about 1000 Hz, and adjust the output voltage of the oscillator to obtain a low sound. In the *collector circuit* now remove the voltmeter V, and replace it by the earphones used in the base circuit, reconnecting the base circuit so that the oscillator provides an input. Compare the loudness of the sound now heard in the collector circuit with the low sound heard previously in the base circuit. Investigate the effect on the loudness when $V_c$ is varied from zero to about 6 V, and comment on the result.

3. Investigate the effect of temperature rise of a transistor on the collector current, for a given base current, for example, by holding the transistor in the hand.

195

# EXPERIMENT 81

## Transistor as Amplifier and Switch

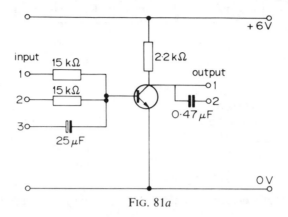

FIG. 81a

APPARATUS

Two *n–p–n* transistors (e.g. ZTX 300) each in a circuit with components as in Fig. 81a (obtainable as a 'basic unit' from most suppliers as Nuffield Advanced Physics item 1075C); 6 V d.c. supply (e.g. dry cells); radio potentiometer 5 kΩ; two voltmeters, 0–6 V; crystal microphone; oscilloscope; leads. *Optional:* two lamp indicator units (as Nuffield item 1075D).

## A. TRANSISTOR AS AMPLIFIER

METHOD

1. The circuit in Fig. 81a can be used as an amplifier. To show that a small change in input voltage can cause a larger change in output voltage, use only input 1 and output 1. Arrange the 5 kΩ potentiometer to vary the input voltage from 0–6 V, with a voltmeter to measure the input voltage, as shown by the heavy additional lines in Fig. 81b. Connect a voltmeter to measure the output voltage.

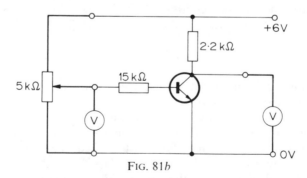

FIG. 81b

2. Starting from zero, vary the input voltage in suitable steps such as 0·2 V and note each time the output voltage. Do this for the full range of input voltage. Observe that when the input voltage rises, the output voltage falls.

3. Now set the input voltage so that the output voltage is about half-way between its largest and smallest values (about 3 V). Connect a microphone between input 3 (Fig. 81a) (which has a capacitor to prevent the microphone altering the d.c. voltage) and 0 V, and an oscilloscope across output 1 and 0 V. Whistle softly, or speak softly, into the microphone and observe the amplified waveform on the oscilloscope.

MEASUREMENTS AND GRAPH

| Input voltage /V | | | | | | |
|---|---|---|---|---|---|---|
| Output voltage /V | | | | | | |

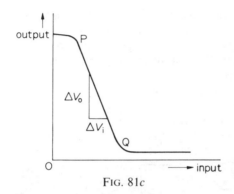

FIG. 81c

Plot a graph of output voltage $V_o$ against input voltage $V_i$ (Fig. 81c).

Observe that the output voltage falls sharply along a straight line over a narrow range of input voltage. Measure the ratio $\Delta V_o / \Delta V_i$, $\beta$, from the slope of the line PQ:

$$\frac{\Delta V_o}{\Delta V_i} = \;.. \;= \text{voltage amplification factor}$$

*Question.* From your graph, why is about 3 V the best value for setting the output voltage in the microphone experiment in 3?

*(Experiment continued overleaf)*

197

## 81. (*Continued*)

### B. TRANSISTOR AS SWITCH

PRINCIPLE

The characteristic in Fig. 81*c* shows that (i) when the input is 0 V or *off*, the output is 6 V or *on*, (ii) when the input is 6 V or *on*, the output is near 0 V or *off*. This property of the characteristic can be used to make a 'transistor switch'. As we shall show, it can also be used to make a transistor switch another transistor on or off in circuits.

METHOD

*Bistable circuit*

1. Use two of the circuits (basic units) in Fig. 81*a*. Connect the output 1 of the first (transistor TR1) to input 1 of the second (transistor TR2) (Fig. 81*d*). Connect a voltmeter V (or lamp indicator unit) across output 1 of each of the two circuits. Connect input 1 of the first circuit to +6 V and observe the output of the second circuit. The output of the first circuit falls practically to 0 V (see Fig. 81*c*) so the output of the second circuit rises to +6 V.

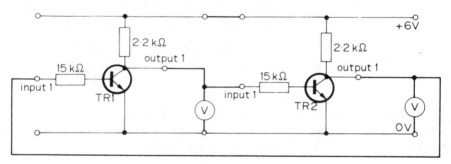

FIG. 81*d*

2. Now connect output 1 of the second circuit to input 1 of the first, forming a 'feedback loop' as in Fig. 81*d*. The voltage fed back keeps the circuit in a stable state. But it has two stable states, either with the first circuit *on* and the second *off*, or the other way round. To switch from one state to the other, touch a lead, for a moment, from +6 V to the input of the transistor whose output is then *on*.

This bistable circuit 'remembers' what last happened; it is a basic memory device used in computers and pocket calculators.

*Astable circuit*

1. Connect an oscilloscope across output 1 and 0 V of one basic unit circuit (Fig. 81*a*); use the oscilloscope switched to indicate d.c. voltages at a sensitivity of about 2 V cm$^{-1}$. Now connect input 1 to +6 V, so that the output is at 0 V.

2. Connect input 3 (with the 25 $\mu$F capacitor, see Fig. 81*a*) first to +6 V and then to 0 V and watch the oscilloscope as you do so. As the connection is made, the output rises to +6 V, but after a short time it falls to 0 V again because the 25 $\mu$F capacitor charges the 15 k$\Omega$ resistor and the voltage input to the transistor rises again.

So if input 3 of this circuit arrangement falls to 0 V, the output rises to +6 V for a short time and then falls back again to 0 V. This brief or *pulse* voltage can be used to make a second circuit behave in the same way.

3. Connect input 1 of TR1 and TR2 in two basic units to $+6$ V (see Fig. 81$e$). Connect output 1 of TR1 to input 3 of TR2, and output 1 of TR2 to input 3 of TR1 (another feedback loop). Observe

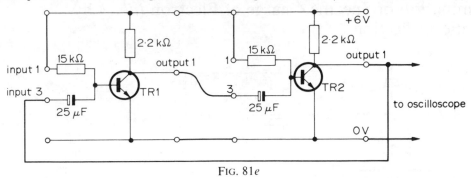

FIG. 81$e$

the trace on an oscilloscope connected across output 1 and 0 V of either circuit. The output rises and falls regularly as each circuit sends a pulse to the other. Lamp units connected to each circuit will flash on and off alternately. The whole arrangement is an 'astable' circuit, which provides pulses of voltage.

4. The pulse rate of the circuit depends on the value of the circuit capacitance. Reconnect the circuit but this time use output 2 of each circuit—see Fig. 81$a$—instead of output 1. The 0·47 $\mu$F series capacitor reduces the capacitance to about 0·47 $\mu$F, so that the capacitors now charge up in a few milliseconds. The pulse rate is thus in the audible frequency range. The output can be seen on an oscilloscope, or heard on a high-impedance earpiece, or amplified and heard on a loudspeaker.

This circuit is used to provide clock pulses for many purposes, including communication and computation.

# EXPERIMENT 82

# Determination of Specific Charge of Electron $(e/m_e)$ by Magnetic Deflection

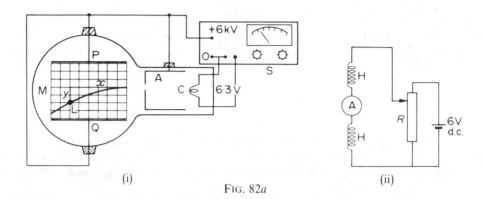

FIG. 82a

## APPARATUS

TELTRON tube (comprising electron gun—cathode C and anode A—fluorescent graduated scale M and plates P and Q); Helmholtz coils H, H; EHT supply 0–6 kV, low-voltage supply 6·3 V a.c.; d.c. supply 6 V such as 3 accumulators; rheostat about 15 Ω, 2 A; meter 0–1 A.

### SAFETY PRECAUTIONS

*Do not* change circuit connections with the power supply switched on. Always reduce the voltage to zero and *switch off*; it may be necessary to wait some time after switching off to allow the smoothing capacitors in the power supply to discharge completely.

## METHOD

1. Arrange the required electrical circuits in Fig. 82a (*i*) and (*ii*). These are:

(*a*) heater or cathode supply of 6·3 V, anode A with voltage of 0–6 kV,
(*b*) plates P and Q, *connected together*, with the same voltage as A,
(*c*) current for Helmholtz coils H, H measured by ammeter A, supplied by potential divider using R across the 6 V d.c. supply, as shown in Fig. 82a (*ii*).

Note that the zero terminal of the EHT supply is connected to one side of the heater or cathode C.

2. Before starting the experiment, record the number of turns $N$ of the Helmholtz coils and their radius $r_H$.

3. Switch on the heater supply of 6·3 V, then the anode voltage $V$ of say 2 kV (this is also the common voltage of the plates P and Q). A horizontal luminous beam should be seen on the graduated screen M along the x-axis. Now switch on the circuit supplying current for the Helmholtz coils. The beam should be deflected along a circular arc.

Adjust the current in the coils until the circular trace passes through a point such as L near the end of the graduated screen M whose x-coordinate (the horizontal distance from the origin) and y-coordinate (the vertical distance from the x-axis) can both be accurately read, for example, $x = 7·0$ cm, $y = 1·0$ cm. Record the x and y values and the current $I$. By altering the rheostat in the potential divider circuit, obtain four other values of current $I$ and four sets of x and y values—each time adjust the current so that the x- and y-coordinates can be accurately read, as above.

200

MEASUREMENTS

$V = \ldots$ V
$r_H = \ldots$ m
$N = \ldots$

| $I/A$ | $x/mm$ | $y/mm$ | $r = \dfrac{x^2 + y^2}{2y}$ /m | $\dfrac{1}{r}$/m$^{-1}$ |
|---|---|---|---|---|
|  |  |  |  |  |

CALCULATION

If $V$ is the anode voltage and $v$ is the velocity of the electron of mass $m_e$ and charge $e$, then

$$\tfrac{1}{2}m_e v^2 = eV \quad \text{or} \quad v^2 = 2\frac{e}{m_e}V \quad \ldots\ldots\ldots\ldots \quad (1)$$

Also,

$$Bev = \frac{m_e v^2}{r} \quad \text{or} \quad v = Br\frac{e}{m_e} \quad \ldots\ldots\ldots\ldots \quad (2)$$

From (1) and (2),

$$\frac{e}{m_e} = \frac{2V}{B^2 r^2} \quad \ldots\ldots\ldots\ldots \quad (3)$$

(i) If $N$, $r_H$ and $I$ are respectively the number of turns, radius and current in the Helmholtz coils, the field value $B$ is given approximately by, if $\mu_0 = 4\pi \times 10^{-7}$ H m$^{-1}$,

$$B = \frac{0.72\mu_0 NI}{r_H} \quad \ldots\ldots\ldots\ldots \quad (4)$$

(ii) The radius $r$ of the circular arc produced by magnetic deflection is given by

$$r = \frac{x^2 + y^2}{2y} \quad \ldots\ldots\ldots\ldots \quad (5)$$

Calculate the different values of $r$ obtained and enter the results in the table of measurements. Obtain the values of $1/r$ and enter the results.

(*Experiment continued overleaf*)

## 82. (*Continued*)

Plot a graph of $1/r$, the inverse of the radius or curvature of the arc, against $I$, the current. Draw the best straight line passing through the origin. Measure the gradient $g(=a/b)$ of the line (Fig. 82b).

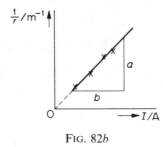

FIG. 82b

From (3) and (4),

$$\frac{e}{m_c} = \frac{2Vr_H{}^2}{0.72^2\mu_0{}^2N^2}\left(\frac{1/r}{I}\right)^2 = \frac{2Vr_H{}^2}{0.72^2\mu_0{}^2N^2}g^2$$

Substitute the values of $V$, $N$, $r_H$, $\mu_0$ and $g$ in this expression and calculate the value of $e/m_c$.

### CONCLUSION

The specific charge of the electron, $e/m_e$, was found to be ... C kg$^{-1}$.

### ERRORS

1. The gradient $g$ of the straight-line graph is subject to error. Since $g^2$ is required in the expression for $e/m_c$, a percentage error of say 5% in $g$ produces a percentage error of 10% in $e/m_c$. From the origin, draw (i) the line of least slope through the points and (ii) the line of greatest slope and find the error in $g$ and so the percentage error in $g$.

2. The voltmeter of the EHT supply may read incorrectly, so that $V$ is in error. The ammeter A may be subject to error.

3. The electron beam has some thickness and this introduces error in setting the beam to pass through a definite point on the graduated scale.

In Sir J. J. Thomson's classical method for finding $e/m_c$, a p.d. $V'$ say was applied between the plates P and Q. This produced an electric field of intensity $E(=V'/d$, where $d$ is the separation of the plates) perpendicular to the magnetic field $B$ which could *neutralise* the deflection due to $B$ alone. The beam then passed horizontally through P and Q. In this case, $Ee = Bev$, where $v$ is the velocity of the electrons, so that $v = E/B$. Thus $v$ can be calculated and the result used in equation (2) to find $e/m_c$. (The use of equation (1) for $v$ *assumes* that the electrons are emitted from the cathode with zero velocity.)

Using the EHT supply, you should apply a p.d. between the plates P and Q to neutralise the magnetic deflection and observe that the electron beam becomes roughly horizontal. In the TELTRON tube, P and Q are so far apart that the electric field becomes non-uniform at the edges, leading to considerable error in the result for $v$ if it is calculated from $v = E/B$.

## QUESTIONS

1. What is the purpose of connecting P and Q together at the same potential? Could the potential of P and Q be different from that of A in this experiment?

2. Does it matter which way round the Helmholtz coils are connected to each other to produce the magnetic deflection? Explain your answer. Draw a rough sketch of the field variation between the two coils.

3. In your opinion, what is the most serious error in this experiment? Give a reason for your answer.

4. Instead of using accumulators for the 6 V d.c. supply for the Helmholtz coils, the 6 V supply is obtained from a rectified a.c. supply. The electron beam seen on the screen is then spread into a fan-shaped circular arc. Explain this appearance of the beam.

# EXPERIMENT 83

# Measurement of Charge on Electron—Millikan's Method

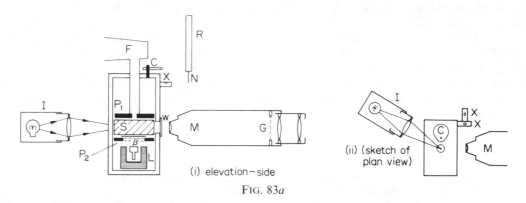

(i) elevation—side

(ii) (sketch of plan view)

FIG. 83a

APPARATUS

Suitable apparatus, which differs in detail from that illustrated in Fig. 83a, is available from most major suppliers (see their instructions for methods of illumination and spraying of oil drops into the cell).

In addition: 0–1000 V power pack, well smoothed, with coarse and fine potentiometer control of voltage, three-position switch for reversing the polarity or shorting of the plates; voltmeter of suitable range; low-voltage supply for illuminating lamp; oil spray (metal nozzle) and vacuum oil; weak sealed $\beta$ source, 1 $\mu$C; stop-watch; transparent mm scale object to calibrate eyepiece graticule; fine Cu–constantan thermocouple, beaker, thermometer in $\frac{1}{10}$ °C, means of measuring spacing (see 7 below).

METHOD

1. *Focusing and adjustment of illumination.* The Millikan cell with parallel plate electrodes $P_1$, $P_2$, microscope and illuminant are mounted on a common base (Figs. 83a(i), (ii)). Place the rod R through the aperture in the top of the cell casing so that the needle N at its tip passes through the pinhole in $P_1$. Switch on the illuminant I and focus the microscope M on the tip of the needle. Swing the lamp and lens I, which moves about an axis through the pinhole, until at an angle of about 15° to the microscope axis, one edge of N is brightly illuminated and there is still sufficient scattered light to illuminate the mm scale on the eyepiece graticule G of the microscope. It is convenient to work in a dimly lit part of the laboratory. Focus I if necessary.

2. *Levelling.* Use the levelling screws on the base of the apparatus to level the apparatus in both perpendicular horizontal directions by means of the spirit levels X, X. Final levelling is obtained later.

3. *Obtaining oil drops.* Replace the needle and rod R by the funnel F, whose tube rests over the hole in $P_1$. Connect $P_1$ and $P_2$ to the power pack, the upper plate $P_1$ being joined to the *earthed* output connection of the pack. Connect the voltmeter to read the potential difference applied to the plates. Set the voltage at about 200 V, but set the switch on the power pack so that the plates are short-circuited.

Squeeze the oil spray, and check that a fine mist of droplets emerges. Spray droplets into F, when they should appear in the field of view of M as bright 'stars' against the background light. Repeat the adjustment of the illumination if droplets are not observed, or the contrast is too poor for comfortable observation. The droplets, falling under gravity, appear to rise in the inverted image seen through M. If the lamp is of too high a power there may be convection currents which reverse or disturb this motion. Such excessive heating of the cell must be avoided.

With a cloud of droplets in view in M, operate the power-pack switch, first one way and then the other, making the upper plate alternately positive and negative. Note that both positively and negatively charged droplets are present, and that the magnitudes of their charges vary, so that those droplets which are attracted to the upper plate may reverse their motion, be slowed down in their fall under gravity or remain more or less stationary when the field is applied.

4. *Control of drops and final levelling.* Obtaining further clouds of drops as necessary, practise selecting a drop that falls fairly slowly under gravity and can be balanced by a p.d. of some 200 V, and holding it balanced in the centre of the field of view. Use the coarse and fine controls and obtain experience in moving and stopping the drop at will.

When a drop has been selected and is roughly balanced remove F and swing the cover C to close the aperture in the casing, so that the drops are not disturbed by movements of air in the room. If the previous levelling was not exact the electric force $q.E$ on the drop $d$ will have a sideways resultant with the weight $mg$ as Fig. 83$b$. The drop may thus move laterally in the field of view, or may move towards or away from the microscope, so that continual re-focusing is needed. Note the directions in which the drop is moving (remembering the reversal of the image for lateral drift), and use the levelling screws to correct the orientation of the plates, as Fig. 83$b$.

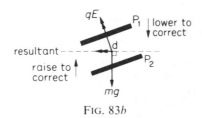

FIG. 83$b$

5. *Measurement of balancing potentials.* Obtain a balanced drop as above, and balance it with care, using the fine control. Observe the drop at intervals of about a minute, looking for slow vertical motion. When the drop is stationary, record the balancing potential $V$. If the $\beta$-source is provided, slide it in its lead 'castle' L (Fig. 83$a$(i)), under the central portion of the lower plate $P_2$, which has an aperture covered with a conducting foil thin enough to admit $\beta$ particles to the cell. Short-circuit the plates, and allow the drop to fall freely towards the lower plate, near which ions of sign opposite to the charge on the drop may be expected to be found (why?). Re-apply the balancing potential, increasing it as necessary, to bring the drop back to the centre of the field of view. Repeat if the (rough) balancing potential is unchanged. If it has altered slide the $\beta$ source to the end of the casing and obtain an accurate balance as before. Record the new balancing potential. Repeat this process until the drop has lost so much charge that the balancing potential cannot be supplied by the power pack and the drop is lost. Usually the potential will increase, but some ions of similar charge may attach themselves to the drop, increasing its charge and providing a check on previously obtained balances. A series of at least five or six different accurately found balance potentials is required for a satisfactory result. If this is too difficult, make one balance voltage observation on each of many drops.

6. *Timing of free fall.* Increase the applied p.d. to raise the drop, switch off so that the plates are short-circuited, and time the free fall of the drop over a noted number of graticule divisions over the middle third of the field of view.

7. *Final measurements.* (1) Use a fine copper–constantan thermocouple through the pinhole in $P_1$ to record the temperature inside the cell. The other junction may be immersed in a beaker of water whose temperature is adjusted for no thermocouple voltage, and measured.

(2) Record the spacing $d$ between the plates accurately, using a travelling microscope, or by measuring the width of the insulated spacers S with a micrometer.

(*Experiment continued overleaf*)

## 83. (*Continued*)

(3) Focus the microscope on a ruled mm scale object, and determine the actual distance corresponding to a noted number of divisions on the graticule as seen through the eyepiece.

(4) Record the density of the oil.

MEASUREMENTS

| Balancing voltage $V$ /V | | | | | | |
|---|---|---|---|---|---|---|
| Nearest integer $n$ | | | | | | |
| $V.n$/V | | | | | | |
| Average $V.n$/V | | | | | | |

| Time of fall /s | | | |
|---|---|---|---|
| Graticule divisions | | | |
| Distance /mm | | | |
| Velocity $v$/m s$^{-1}$ | | | |

*Graticule calibration.* . . . divs. on the graticule correspond to . . mm at the object plane of the microscope.

$$\text{Plate spacing } d \quad = \text{ .. mm} = \text{ .. m}$$
$$\text{Temperature of air in cell} = \text{ .. °C}$$
$$\text{Density of oil} \quad = \text{ .. kg m}^{-3}$$
$$\text{Viscosity of air} \quad = \text{ .. N s m}^{-2} \text{ (from tables)}$$

CALCULATION

1. *Discrete unit of charge.* When a drop of mass $m$ is balanced by a p.d. $V$, the plate spacing being $d$, and the charge on the drop $q$:

$$\frac{qV}{d} = \text{Apparent weight of drop}$$

since the field between the plates $= V/d$. For a drop of constant mass, if the charge $q$ is an integral multiple $q = n.e$ of some discrete unit of charge $e$,

$$\frac{neV}{d} = \text{Apparent weight of drop} = \text{constant}$$

If enough values of $V$ for one drop are available, find integers $n$ such that the product $V.n$ is a constant within the limits of experimental error. Record the values of $n$ and of $V.n$ in the table (see also NOTES).

2. *Drop radius.* Convert the distances of free fall to metre and calculate the velocity of fall $v$ in m s$^{-1}$ for each determination. Applying Stokes' law,

$$\tfrac{4}{3}\pi a^3(\rho - \sigma)g = 6\pi\eta a v$$

where $a$ is the drop radius in metre, $\rho$, $\sigma$ the densities of oil and air respectively in kg m$^{-3}$, and $\eta$

206

the viscosity of air in $N s m^{-2}$ at the temperature of the experiment. Calculate the radius from:

$$a^2 = \frac{9 \cdot \eta \cdot v}{2g(\rho - \sigma)} = \ .. \ m^2; \quad a = \ .. \ m$$

3. *Electronic charge.* From the second equation the constant value of $V \cdot n$ is $d/e$ times the apparent weight of the drop. Thus

$$e = \frac{d}{V \cdot n} \cdot \frac{4}{3} \pi a^3 (\rho - \sigma) g$$

Substitute the value of $a$ found above and the averaged constant value of $V \cdot n$, giving:

$$e = \ .. \ coulomb$$

If too few values of $V$ are available for each drop to identify integers, calculate the charge $q = n \cdot e$ on each drop, using the above equation. Results for several drops will be required. Tabulate the values of $q$, and identify integers $n$ such that

$$\frac{q}{n} = constant = e \ coulomb$$

CONCLUSION

The .. values of $n$ provide evidence that electric charges occur in multiples only of a constant elementary charge, $e$, $= \ .. \ C.$

ERRORS

1. The values of $e$ obtained are likely to show a systematic variation with the drop radius $a$ used, being higher the smaller $a$ becomes. The effect is due to departures from Stokes' law, and a correction may be found in standard textbooks. Alternatively, plot the values of $e$ against $1/a$ and draw a smooth curve (which is not likely to depart far from a straight line) through the points. Read the corrected value of $e$ from the intercept where $1/a = 0$.

2. Why is it not possible to identify large integers with certainty?

3. From times of fall of one drop measured at intervals, how would you obtain evidence that a drop is, or is not, evaporating?

NOTES

1. The radioactive source mentioned provides ions in the cell so that the chance of a change of charge is increased. Without the source, changes may be infrequent. It may then be best to obtain one balance voltage for each drop, and follow the alternative calculation in section 3.

2. If integers are not obvious by inspection, they may be found by using a slide rule. For $V \cdot n =$ constant, set the rule to divide the largest value of $V$ by the reciprocal of a small 'test' integer. Try various integers, until the reciprocals of other integers on the sliding scale appear opposite measured values of $V$ on the fixed scale. For $q/n = $ constant, simply divide values of $q$, starting with the smallest, by integers until an integer on the sliding scale corresponds with each value of $q$, as above.

# EXPERIMENT 84

# Investigation of Geiger–Müller Tube and Scaler

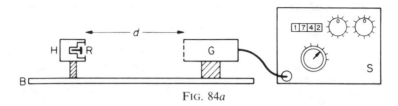

FIG. 84*a*

APPARATUS

End window Geiger–Müller (G–M) tube G of $\beta$, $\gamma$ sensitive type, e.g. Mullard $MX$108, $MX$168, 20th Century $EWG$5H; decade scaler S with variable EHT supply; sealed radioactive source R, $\beta$-active (e.g. $^{90}$Sr) or $\gamma$-active (e.g. $^{60}$Co); source handling tongs; some form of 'optical bench' B with source holder H; stop-watch.

## 1. GEIGER–MÜLLER TUBE CHARACTERISTIC

METHOD

Clamp the G–M tube G to one end of the optical bench B, and connect its coaxial cable to the input socket of the scaler S. Place the radioactive source R in a holder H, about 20 cm from the G–M tube. Switch on the scaler, with the variable EHT voltage set at a minimum, and allow it to warm up for a few minutes.

The scaler controls should include a switch which enables counting to be started or stopped, and a test facility which provides for the scaler to count regular pulses (50, 100 or 1000 per second) generated internally. Switch to *test* and then to *count* and check that the counter operates. Stop the scaler counting, reset the registers to zero, and use a stop-watch to record the number of counts from the test pulses in several 100-second periods. Check that this rate of counting agrees with the scaler specification. Practise using the stop-watch and count switch until you can obtain reasonably consistent results.

Set the scaler to count pulses from the G–M tube, and slowly raise the EHT voltage applied to the tube until the scaler just begins to count. Keeping the source in the same position, record the count obtained in periods of at least 60 or 100 seconds for this voltage and for a series of higher voltages. If the rate of counting at higher voltages begins to rise after remaining much the same for a range of voltage, *do not* raise the voltage any higher, or the tube may suffer damage.

MEASUREMENTS

| EHT voltage, $V$/V | | | | | | |
|---|---|---|---|---|---|---|
| Count in .. second | | | | | | |
| Count-rate, $C$/s$^{-1}$ | | | | | | |

## GRAPH

Plot a graph of the count-rate $C$ v. EHT voltage $V$ (Fig. 84$b$). Record the voltages $V_1$ and $V_2$ between which the rate of counting does not change very much.

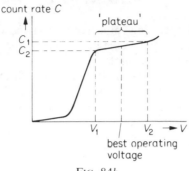

FIG. 84$b$

Plateau starts at $V_1 = $ .. V
Plateau stops at $V_2 = $ .. V

Between $V_1$ and $V_2$ the count-rate changes from $C_1 = $ .. to $C_2 = $ .. counts per second.

## DISCUSSION

Is any particular value of the counting-rate the 'correct' one? (Note that higher rates at high voltages may be due to spontaneous discharges inside the G–M tube not initiated by incoming radiation.) In what region is the counting-rate very likely to be incorrect?

The voltage applied to the tube may well vary during the experiment. If, from the graph, the count-rate varied from $C_1$ to $C_2$ as the voltage rose from $V_1$ to $V_2$, work out how much the rate would change if the voltage rose or fell by, say, 10 V from some setting in the middle of the 'plateau'. What percentage error would such a voltage variation introduce into the counting-rate? Why is the voltage in the centre of the plateau labelled 'best operating voltage' in Fig. 84$b$?

## CONCLUSIONS

For reliable comparisons of rates of counting, the G–M tube should be operated at about .. V.
Percentage error in count rate due to 10 V change in voltage = .. %.

*(Experiment continued overleaf)*

# 84. (*Continued*)

## 2. THE COUNTING OF RANDOM PULSES AND BACKGROUND RADIATION

Set the EHT voltage to the best operating value (see part 1) and watch the counter with the source removed. Note that occasional pulses still occur; this is due to the so-called 'background' radiation. Observe whether pulses occur regularly, or not.

Since the background radiation is present in all experiments, it will have to be measured and subtracted from experimental readings for these to be reliable. Its measurement also illustrates another important feature of the counting of pulses from radioactive sources.

Record the number of counts from the background in as many as 100 or more 10-second intervals. Tabulate the number of times each count was obtained, as below.

### MEASUREMENTS

| Counts in 10 second | | | | | | | | |
|---|---|---|---|---|---|---|---|---|
| No. of times obtained | | | | | | | | |

### GRAPH

Plot a graph of the number of times each count was obtained against the values recorded, as Fig. 84c.

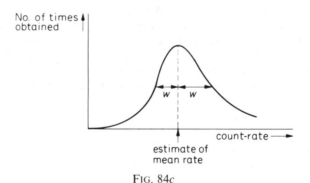

FIG. 84c

Is it possible that this spread of results could be due to errors in timing the 10-second intervals?

### CALCULATION

Find the mean count in 10 second from (i) the maximum of the curve, (ii) the total count recorded divided by the total number of 10-second intervals. (Could these values disagree?)

Find the average width $w$ of the curve (see Fig. 84c) on each side of the vertical through the maximum, at which the number of times obtained is half the maximum value. Assuming that there is a fair chance that any one result will be within $\pm w$ of the middle or mean value, estimate the percentage error in one 10-second reading of the background radiation, which is $(\pm w/\text{mean value}) \times 100$ per cent.

## THEORY

The curve obtained in Fig. 84c arises from the random nature of the background radiation. Since all radioactive sources emit randomly, the same problem arises in all such measurements.

For a fairly large mean count $n$ in any one interval, the width $w$ of the corresponding curve of results, or *deviation*, may be shown by statistics to be roughly equal to $\sqrt{n}$, assuming there is a fixed chance of a pulse occurring in any short interval of time. Thus the deviation ($\sqrt{n}$) increases with $n$ but the *percentage deviation* ($\sqrt{n}/n = 1/\sqrt{n}$) *decreases* with $n$.

## CONCLUSIONS

1. The mean background radiation was .. counts in 10 second
$$= \text{.. counts/second.}$$

2. The percentage deviation in *one* 10-second reading was:

(1) from the graph,
$$\frac{w}{\text{mean value}} \times 100\% = \text{.. } \%$$

(2) from theory,
$$\frac{1}{\sqrt{\text{mean count in 10 s}}} \times 100\% = \text{.. } \%$$

3. The percentage deviation in the value obtained from a total of .. counts in .. 10-second intervals
$$= \frac{1}{\sqrt{\text{total count}}} \times 100\% = \text{.. } \%$$

4. To achieve a certain percentage accuracy in measuring any rate of counting, the total count must be such that ...

## PROBLEMS

1. Measure the count rate due to a luminous watch in front of the G–M tube to within 5%.

2. What would you consider convincing evidence that the radiation from a source is emitted randomly? Find whether the number of counts obtained using a watch or one of the sources near the counter varies from measurement to measurement in the way you would expect if the decay is random.

# EXPERIMENT 85

# Investigation of Inverse Square Law for $\gamma$-rays

### SAFETY PRECAUTIONS

*Always* handle the radioactive source with tongs, and never place it near any part of the body.
 *Keep the source* in a shielded container when not in use.

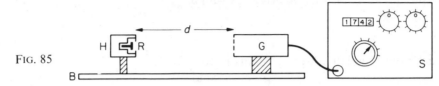

FIG. 85

### APPARATUS

Geiger–Müller (G–M) tube G sensitive to $\gamma$-radiation; decade scaler S with EHT supply; sealed gamma source, e.g. $^{60}$Co; source handling tongs, source holder H, simple form of 'optical bench' to hold source and G–M tube; stop-watch, ruler.

### METHOD

1. Connect the G–M tube G to the scaler, and mount it at one end of the optical bench. Set the EHT voltage to a value in the middle of the tube 'plateau' (see Expt. 84, p. 209) and switch on. Use the test facility (if provided) to check that the scaler is counting correctly.

2. Fix the $\gamma$-active source R in a holder H facing the G–M tube window, and move it until the count-rate, measured roughly over a few seconds, approaches the highest rate at which the scaler can count. Measure the distance $d$ cm from the tube window to a convenient reference point on the source holder. Record the count obtained over a time sufficiently long to attain a percentage uncertainty in the total count of, say, 2%. (See Expt. 84, p. 211.)

3. Move the source back until the count-rate has fallen to a low value, of the order of 10 counts per second. Then make observations of the count-rate as before at several evenly spaced distances $d$ between these two limits and record the results. Work out a suitable time over which to record each count, having first measured the rate roughly at each distance, so that the same degree of accuracy (e.g. 2%) is maintained. Explain why these times become long as the count-rate decreases.

4. Remove the source, placing it in its shielded container, and measure the count-rate due to background radiation. Assuming that the background gives a rate of the order of one per second, which will be subtracted from the values obtained with the source present, explain why an accuracy of about 20% in this measurement will be sufficient.

### MEASUREMENTS

*Background radiation.* Count in .. seconds = .., ..

$\therefore$ background rate $C_1 = $ .. counts per second

| *Type of radiation* = .. | Distance $d$/m | | | | | | |
|---|---|---|---|---|---|---|---|
| | Count | | | | | | |
| | Time /s | | | | | | |
| | Count-rate $C_2$/s$^{-1}$ | | | | | | |
| | Corrected count-rate $C$/s$^{-1}$ | | | | | | |

## CALCULATION

Subtract the background count-rate from the measured values of the rate using the source, and enter the count-rate $C$ due to the source only in the table of measurements. Are there any measurements where this is unnecessary, in view of the accuracy achieved?

## GRAPHS TO TEST FOR INVERSE SQUARE LAW

Plot graphs of (i) $1/d^2$ v. $C$, (ii) $1/\sqrt{C}$ v. $d$. Test whether each set of plotted points is consistent with a straight line, and note whether such a line passes through the origin or not.

Remembering that the reference point used on the source holder may not coincide with the active surface of the source, and that the effective counting space inside the G–M tube may not be close to the window, discuss the effect of such a systematic error in the values of $d$ on the two graphs. Consider whether your graphs offer evidence that the $\gamma$-radiation does, or does not, follow an inverse law. See NOTE 1.

## CONCLUSIONS

The graph of .. v. ... for $\gamma$-radiation is (is not) consistent with an inverse square law. This graph was chosen because ...

## NOTE

1. If an inverse square law is obeyed, the intensity $I \propto 1/r^2$, where $r$ is the distance from the source to the detector. If $d_0$ is the distance to be added to the measured distance $d$ because (i) the reference point on the holder does not coincide with the source, and (ii) the effective counting space inside the G–M tube may not be close to the window, then $r = d + d_0$. Since $I \propto C$,

$$C \propto \frac{1}{(d+d_0)^2} \quad \cdot \quad \cdot \quad \cdot \quad \cdot \quad \cdot \quad \cdot \quad \cdot \quad \cdot \quad \cdot \quad \text{(i)}$$

Rearranging,
$$\therefore \ d + d_0 \propto \frac{1}{\sqrt{C}}. \quad \cdot \quad \cdot \quad \cdot \quad \cdot \quad \cdot \quad \cdot \quad \text{(ii)}$$

From (i) and (ii), what will be the effect of the error $d_0$ on the two graphs plotted if an inverse square law is followed?

2. As a consequence of the random nature of the emission of radiation from radioactive materials, the percentage error of any one total count $N$, where $N = C \times$ time of observation, is given by:

$$\text{percentage error} = \frac{1}{\sqrt{N}} \times 100\%$$

213

# EXPERIMENT 86

# Absorption of Beta-particles by Aluminium

## SAFETY PRECAUTIONS

*Always* manipulate radioactive sources with long tweezers and *never* handle them or place any part of the body in front of them.

*Keep the source* in its container except when it is being used and *replace it* whenever it is not in use.

## APPARATUS

Geiger–Müller (G–M) tube G sensitive to $\beta$-radiation, e.g. MX168; decade scaler S with EHT supply; sealed radioactive beta source R, strontium-90; source handling tongs, source holder H; vernier calipers; aluminium sheets of varying thickness from about 0·5 mm to 5 mm; stop-watch.

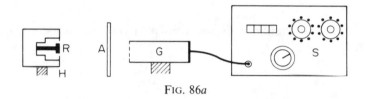

FIG. 86a

## METHOD

1. Connect the G–M tube G to the scaler S (Fig. 86a). Set the EHT voltage to the required value and switch on. Use the test facility (if provided) to check that the scaler is counting correctly, having warmed up.

2. Support the G–M tube at a convenient height in a clamp stand (clamp the holder, *not* the tube, which is fragile). Place the beta-source in another clamp at the same height and about 10 cm from the G–M window. Move the source so that, with the thinnest aluminium sheet, the count-rate is high and measurable on the scaler concerned, for example, of the order of 60–80 per second.

3. Support the thinnest sheet of aluminium A in a wooden clamp between the source and G–M tube. Using the stop-watch, record the count obtained over a time sufficiently long to give a percentage uncertainty in the total count of say 2% (see Expt. 84, p. 211, and *Note* (d) on order of accuracy). Measure the thickness of the aluminium sheet with vernier calipers.

4. Take a series of readings of the counts from the thinnest to the thickest aluminium; roughly aim at the same number of counts to provide the same accuracy. Record the corresponding aluminium thickness after each count.

5. Remove the source and place it in its shielded container. Now measure the count-rate due to background radiation by recording the counts in as many minutes as possible.

## MEASUREMENTS

*Background radiation.* Count in . . seconds = . .

So                          background rate $C_1 = $ . . counts per second

| Aluminium thickness /mm | Count | Time /s | Count-rate $C_2$ /s$^{-1}$ | Corrected count-rate $C$ /s$^{-1}$ |
| --- | --- | --- | --- | --- |
|  |  |  |  |  |

214

## GRAPH

Plot a graph of $\log C$ against the thickness $t$ of aluminium. From your graph: (i) calculate approximately how $\log C$ diminishes with increasing $t$—this is given by the *gradient* of the best straight line through the smaller values of $t$, (ii) record the thickness $t_0$ of aluminium which first reduces the count-rate to a steady minimum value at the horizontal 'tail' of the graph (Fig. 86$b$).

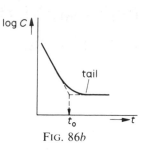

FIG. 86$b$

## CALCULATION

(i) The intensity of radiation $I$ diminishes with small thickness $t$ of aluminium roughly according to the exponential relation $I = I_0 e^{-\mu t}$, where $I_0$ is the incident intensity and $\mu$ is a constant known as the 'linear absorption coefficient'. So

$$\ln I = \ln I_0 - \mu t$$

or

$$\log I = \log I_0 - \frac{\mu}{2 \cdot 3} t$$

if logs to the base 10 are used. From the gradient of the graph, calculate $\mu$:

$$\mu = \ .. \ \text{m}^{-1}$$

Then calculate the *mass absorption coefficient* $\mu_m$ of aluminium from the relation

$$\mu_m = \frac{\mu}{\rho} = \frac{\mu}{2700}$$

where $\mu_m$ is in units $\text{m}^2 \text{kg}^{-1}$ if $\mu$ is expressed in $\text{m}^{-1}$ and the density $\rho$ of aluminium is $2700 \ \text{kg m}^{-3}$; the values of $\mu_m$ enable the relative absorbing powers of different materials to be compared irrespective of their density.

$$\mu_m = \ .. \ \text{m}^2 \text{kg}^{-1}$$

(ii) The thickness $t_0$ of aluminium which reduces the count-rate to a steady minimum value is taken as the 'range' of beta-particles in aluminium. The thickness of an absorber is often expressed in terms of the 'surface density', which is given by

$$\text{surface density} = \text{thickness} \times \text{density of absorber}$$

Calculate the range from

$$t_0 \ (\text{metre}) \times 2700 \ \text{kg m}^{-3} = \ .. \ \text{kg m}^{-2}$$

## CONCLUSION

1. The mass absorption coefficient of aluminium $= \ ..\ \text{m}^2 \text{kg}^{-1}$.
2. The range of beta-particles in aluminium $= \ ..\ \text{kg m}^{-2}$.

## NOTE

(*a*) The count-rate does not reduce to zero as the thickness of aluminium is increased owing to the emission of radiation (X-rays) as the beta-particles (electrons) decrease in energy.

(*b*) The 'surface density' of an absorber is the 'mass per unit area' (and may therefore be expressed in 'milligram per cm²'). Since mass = volume × density = area × thickness × density, mass per unit area = thickness × density.

(*c*) Beta-particles may also be *back-scattered* from the surfaces of metals. Investigate (i) if this is the case, (ii) if so, whether the intensity of the scattered radiation depends on the metal thickness.

(*d*) *Order of accuracy.* If $N$ is the total number of counts, the deviation is $\sqrt{N}$; so the percentage deviation is $(1/\sqrt{N}) \times 100\%$ (see p. 213). Thus for a total count of 10 000, the % deviation is 1%; for a total count of 2500, the % deviation is 2%; and so on.

215

## EXPERIMENT 87

# Half-life of Radon-220, using D.C. Amplifier

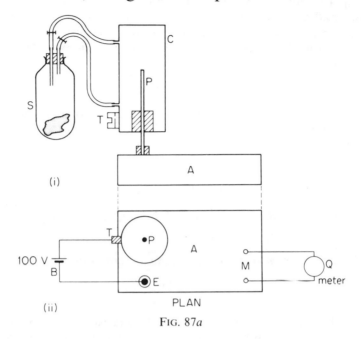

(i)

(ii)

PLAN

FIG. 87a

## APPARATUS

Electrometer and d.c. amplifier, A; ionisation chamber C, with probe P; squeeze bottle S containing thorium hydroxide in a wash leather bag; meter, Q, 0–1 mA or as specified by manufacturer; 0–100 V d.c. power supply; $10^{11}\,\Omega$ input resistor, or $10^{-11}$ A range of electrometer; stop-watch.

## INTRODUCTION

In this experiment, some radioactive gas Radon-220 ('thoron') is injected into an ionisation chamber C (Fig. 87a (i)). As the atoms decay, a current flows between C and the insulated probe P. This current, flowing through a $10^{11}\,\Omega$ resistor, produces a p.d. across it. The electrometer amplifier then produces a much larger current, which is indicated by the meter Q. As the rate of decay of the atoms diminishes, the current falls. The change of activity with time can be followed.

## METHOD

The ionisation chamber C is first fitted on to the electrometer A, with the probe rod P connected to the electrometer input but insulated from C. Having switched on the electrometer, adjust the reading of the output meter Q to zero. Q may be part of the electrometer, or may be connected to its output terminals M. The input electrometer terminal is now connected to earth using the switch provided and Q is then adjusted to read zero.

Insert the $10^{11}\,\Omega$ input resistor, or switch to the $10^{-11}$ A range of the electrometer. Connect a 100 V d.c. supply B between the terminal T of the ionisation chamber casing and E, the earth terminal of the electrometer (Fig. 96a (ii)). The chamber C should be made positive with respect to earth. Unless one is already provided in the ionisation chamber, for safety a 1 MΩ resistor should be connected in series with the h.t. supply. Switch the electrometer to read current. The output meter Q should still show a zero, or nearly zero, reading. If the 100 V d.c. supply is altered, or switched on, the output meter will give a brief reading and then return to zero. Can you explain this?

216

Now connect the squeeze bottle S to the chamber C, as shown. It is safest to have two tubes leading from the bottle to C so that the radioactive gas is pumped round a closed circuit and not into the atmosphere. Unless there is a 1 MΩ safety resistor, it is dangerous to touch the ionisation chamber with the 100 V d.c. supply on.

When the bottle is squeezed, the meter should begin to show a current reading. After one or two squeezes, it should show a full-scale deflection. Stop squeezing, and start a stop-watch when the current on falling passes a convenient mark near the end of the scale. Record the meter reading at convenient intervals, say every 15 seconds.

MEASUREMENTS

| Time $t$ /s | | | | | | | |
|---|---|---|---|---|---|---|---|
| Meter reading $I$ /A | . . | | | | | | |

GRAPHS

(i) Plot a graph of output current $I$ against time $t$ (Fig. 87b (i)).

Find the time, $T_{1/2}$, for the current to fall to half its initial value OL. See whether the time, $T_{1/2}$, for the current to fall from some other value OX to half that value is the same.

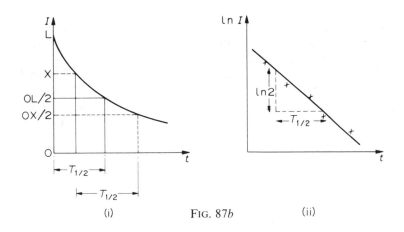

Fig. 87b

(ii) Plot $\log I$ (or $\ln I$) against $t$. (It is convenient to write the smallest currents in units larger than 1, so that the logarithms will all be positive numbers. This makes no difference to the graph other than shifting its zero, since multiplying a number by a fixed factor adds a constant amount to its logarithm.)

Test if the log graph is a straight line. Since $I = I_0 e^{-\lambda t}$, find from the graph a value of the half-life $T_{1/2}$ (Fig. 87b (ii)).

Compare your result with that obtained in (i). Which do you think is the more accurate result. Why?

CONCLUSION

The half-life of Radon-220 is . . seconds.

# EXPERIMENT 88

## Estimation of the Planck Constant

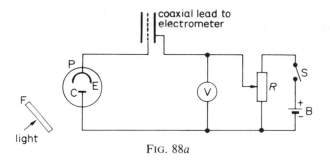

FIG. 88a

### APPARATUS

Vacuum photocell P with potassium emissive surface (e.g. UNILAB), mounted in light-tight box; volt-meter V, 0–2 V; potentiometer $R$, about 1 kΩ; dry battery B; key S; electrometer; coaxial cable to join photocell and electrometer; range of colour filters F; output meter for electrometer, 0–1 mA or 0–100 μA as required; 100 W filament lamp.

### PRINCIPLE OF THE EXPERIMENT

The energy of a photon of light of frequency $v$ is $E = hv$. If these photons eject electrons from a metal surface, the energy of the fastest electrons emitted can be measured by finding the potential difference $V_s$ which just stops them reaching a collector. The energy is given by $eV_s$.

There are two reasons why the emitted electrons do not have energy equal to $E = hv$. One is that a variable amount of energy, up to a value $w_0$ (called the work function of the metal) is used in escaping from the metal. The other is that the emitter and collector must be made of different metals (or the collector would emit electrons too), so they have a small p.d. between them. Together, these effects mean that the electrons have an energy up to, but not exceeding, $hv - w$, where $w$ is a constant taking account of both effects. It is this maximum energy which is measured.

## METHOD

1. The photocell must be mounted in a light-tight box, with an aperture which can be covered by various light filters so that light of different wavelengths can reach the emissive surface E. The box may contain the circuit shown in Fig. 88a. This has a potential divider arrangement formed by the battery B and the potentiometer R, with connections for the voltmeter V and for a cable to connect an electrometer in the circuit in order to measure the photo-electric current.

2. Set up the circuit as shown. Use a coaxial cable to connect the photocell to the input of an electrometer capable of measuring about $10^{-8}$ A. The electrometer actually measures the voltage across its input, in the range 0–1 V. See Expt. 87. A resistance of $10^8 \Omega$ needs to be connected across the input, in which case the p.d. across it indicates a current of the order of magnitude $10^{-8}$ A. In some makes of electrometer this resistance is automatically included on switching to the current range $0$–$10^{-8}$ A; in others, the resistance must be connected in place.

Check that, when white light falls on the cell, the electrometer records a current when the potentiometer R is adjusted so that the stopping potential $V_s$ is zero, as indicated on V. If necessary, use a brighter light source and check the connections.

3. Put a colour filter over the photocell. Illuminate the filter with the light source and observe the photocurrent. Increase the stopping potential $V_s$ by means of R until the current just becomes zero, that is, electrons now do not reach the collector. Record the value of $V_s$ and $\lambda$, the wavelength.

See if there is some current in the opposite direction when the stopping potential is increased further, due to emission from potassium on the collector. If there is an appreciable current, it can be reduced by heating the collector, which is in the form of a wire loop, by passing electric current through it.

Repeat for other colour filters, recording each time $V_s$ and $\lambda$.

## MEASUREMENTS

| Filter colour | Wavelength band passed | | $\dfrac{1}{\lambda_{MIN.}}$ | $\dfrac{1}{\lambda_{MAX.}}$ | Average $1/\lambda$ | Cut-off voltage $V_s$ |
|---|---|---|---|---|---|---|
| | $\lambda_{MIN.}$ | $\lambda_{MAX.}$ | | | | |
| | | | | | | |
| | | | | | | |

Enter values for the limits of wavelength $\lambda_{MIN.}$, $\lambda_{MIN.}$ passed by each filter, from maker's data. Calculate $1/\lambda_{MIN.}$, $1/\lambda_{MAX.}$ and the average of these values, $1/\lambda$.

(*Experiment continued overleaf*)

# 88. (*continued*)

GRAPH

As explained above, the equation (due to Einstein) for the maximum energy of a photoelectron emitted when light of frequency $v$ falls on a metal surface is

$$E = hv - w. \quad \ldots \ldots \ldots \ldots \quad (1)$$

If electrons of charge $e$ are just stopped by a potential difference $V_s$, then

$$E = eV_s = hv - w. \quad \ldots \ldots \ldots \quad (2)$$

If the wavelength passed by the filter is given by $c = f\lambda$, this equation becomes:

$$V_s = \frac{hc}{e} \cdot \frac{1}{\lambda} - k \quad \ldots \ldots \ldots \ldots \quad (3)$$

where $k$ is a constant.

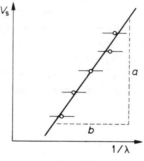

FIG. 88*b*

Plot a graph of the cut-off voltage $V_s$ v. $1/\lambda$ as in Fig. 88*b*, marking the limits of the pass band of each filter on the scale of $1/\lambda$. Test whether the plotted points are consistent with a straight line, and draw in the best straight line. Measure the slope $a/b$ of the line. Draw lines with greater or smaller slope that still fits the points quite well, and note the change in the slope.

220

## CALCULATION

From equation (3), slope of graph $=\dfrac{hc}{e}=\dfrac{a}{b}$

$$\therefore\ h=\frac{e}{c}\times\frac{a}{b}\qquad\qquad\qquad (4)$$

Calculate the value of $h$, using a consistent set of units for $e$, $c$, $a$ and $b$ as indicated below:

| Potential difference ($a$) | 1/wavelength ($b$) | Velocity of light ($c$) | Electron charge ($e$) |
|---|---|---|---|
| V | $m^{-1}$ | $m\,s^{-1}$ | C |

## CONCLUSION

Planck constant, $h$, $=$ .. J s.

## ERRORS

The value for $h$ obtained is likely to have the correct power of 10, but to differ from the accepted value by a substantial factor. Samples of the photocell give differing values, which also vary appreciably with time, and some commercial cells will even give a graph of reversed slope. The emissive surface is not a pure metal, nor is it free from adsorbed gas.

There may well be a current when the p.d. is raised above the 'cut-off', indicating that the collector C has an emissive coating. The 'cut-off' value will then be merely the voltage where a change of brightness raises the reverse current as much as it raises the emitter current, so that the net current is unchanged.

# PART EIGHT

# Short Experiments

The short experiments which follow are each intended to take 10–20 minutes, starting with the apparatus set up and working. You need longer if you have to set it up yourself.

The experiments are designed to help you *observe*, *describe*, and *think about* experimental phenomena and their implications. They suggest some particular arrangement of some apparatus, definite things to do with the apparatus, and questions to answer. The *result* of one of these experiments is, not the value of some quantity or other, but the *ideas* you have about it, mostly written down on paper.

Here are some further brief guidelines. When asked to take a reading, do it carefully, and try to form some idea of how accurately you know the value. You will not be asked to take many readings, but each should be worth having.

When asked to observe what happens, take a good long look and try it more than once. It is easy to miss something important. Also, *think* while you look. When you expect something to happen, it is often easier to see it.

Describing what happens is not as easy as it sounds. You will need scientific words—words like 'resonance', 'phase', or 'potential difference'—and you will need to know how to use them. A good description will explain what is happening, as well as saying what you can see with your eyes.

In some experiments, you will be asked to *suggest* further things to do, and to explain—perhaps defend—your suggestions. Here the hard thing is to be imaginative and far-seeing enough, without letting go of common sense. You will find that this is a rather good test of how well you understand what is going on in the experiment.

## 89. INDUCED VOLTAGES IN A COIL

APPARATUS: 2 small coils of about 250 turns, able to carry 2 A; steel rod (e.g. retort stand rod) about 0·2 m long; mains-driven low-voltage alternating supply, about 6–12 V; cathode-ray oscilloscope, C.R.O.; metre rule; leads; rheostat (about 12 Ω, 5 A; a.c. ammeter 0–5 A.

One coil X is to be connected to the supply, in series with an a.c. ammeter and a rheostat; the supply voltage and rheostat are adjusted so that the ammeter indicates 1 A. The other coil Y is to be connected to the Y-input of the oscilloscope, which is adjusted initially to give a trace about 20 mm peak to trough with the coils about 50 mm apart.

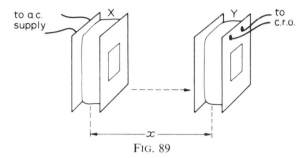

FIG. 89

METHOD

1. Put the two coils facing each other as in Fig. 89, but close together so that they are almost touching. With the current in the coil X joined to the supply at 2·0 A, observe the trace on the C.R.O. and measure the distance from a peak to a trough.

Vertical height of trace = ... mm

2. Use the rheostat to reduce the current in coil X to 1·0 A. Again measure the height from peak to trough of the trace on the C.R.O. Explain this result.

Vertical height of trace with 1·0 A in coil X = ... mm        Explanation ...

3. Insert the steel rod through the coils so that the same length projects from each coil. Observe, without making measurements, what has happened to the ammeter reading. Adjust the rheostat so that the current is again 1·0 A, and describe how you did it. Explain (a) why the current changed; (b) why the change is big or small; and (c) why it can be made as big as it was by altering the supply and rheostat.

Observation of ammeter when steel rod is inserted ...
Description of how to restore the current to 1·0 A ...
Explanation of effects (a) ... (b) ... (c) ...

4. With the steel rod in place, and the current in X at 1·0 A, adjust the C.R.O. gain control to get a reasonable trace. How does the induced voltage in coil Y compare with that obtained in 2? Explain the difference.

Comparison of induced voltage with and without steel rod ...        Explanation ...

5. Use the rheostat to increase the current in coil X to 2 A again. Observe the new height of the C.R.O. trace. Suggest a reason why it *might not* now be twice as large as with 1·0 A in coil X.

Effect on trace ...                Reason why trace may not double in height ...

6. If you alter the distance $x$ between the coils, the induced voltage in coil Y, as shown by the trace on the C.R.O., changes less with the steel rod in place than without the rod. (Try it with the rod in place and then taken out, adjusting the current in each case to the same value.) Suggest a reason for this effect and propose ways of investigating it further.

Reason ...                Proposals for investigation ...

# 90. RESONANCE IN A LOUDSPEAKER

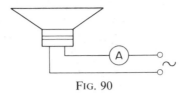

FIG. 90

APPARATUS: cheap loudspeaker, about 100 mm diameter; signal generator with range 10–1000 Hz (or 30–3000 Hz) and low-impedance output (or with suitable amplifier if high-impedance output); multimeter such as Avometer; leads; disc of paper of similar material to loudspeaker and about same size; top-pan balance.

Connect the loudspeaker, resting cone upwards on the bench, to the signal generator low-impedance output, with the Avometer in series. (If the signal generator has only a high-impedance output, connect the loudspeaker to an audio amplifier fed from the signal generator.) Set the Avometer to a.c., and choose a current range and a setting of the gain control of the generator, so that the meter deflects nearly full-scale at the low end of the frequency range such as 30 Hz.

METHOD

1. A loudspeaker, with a meter to measure the alternating current in it, is connected to a signal generator. Starting at the lowest frequency available (10 Hz or 30 Hz), increase the frequency gradually, to about 500 Hz. Observe the loudspeaker cone, listen to the sound from it, and observe the behaviour of the meter. Describe what happens (see below), using suitable terms in your answer such as oscillation, resonance, frequency.

> *Description:*
> Motion of cone …
> Sound produced …
> Current in meter …

2. Locate and record the frequency at which the cone vibrates with maximum amplitude, and the sound is very loud. If the cone has a mass $m$ in kg, and is restrained by springs which need a force $k$ in newton per metre displacement, theory suggests that this frequency $f$ in Hz will be given by

$$f = \frac{1}{2\pi}\sqrt{\frac{k}{m}}$$

Describe *how* you would use small weights which can be put on the cone to displace it, so as to measure $k$ (do not *do* it).

> Frequency of maximum amplitude …
> Suggested method of finding $k$ …

3. Weigh the paper disc provided, and make a *rough* estimate of $m$. Use the equation given above, and the values of $f$ and $m$, to calculate a value of $k$. Explain whether you think the value you get is reasonable or not.

> Estimate of $m$ … kg
> Calculation of $k$ …
> Discussion of value obtained …

4. Recall what happened to the current in the loudspeaker, as the frequency of the current increased from below $f$ to above $f$. (Repeat if necessary.) Suggest what reasons you can for the behaviour of the current.
> Reasons …

## 91. CURRENT IN A CONDUCTOR

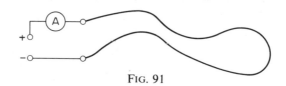

FIG. 91

APPARATUS: 3 metres of 36 s.w.g. enamelled copper wire; low-voltage d.c. supply; ammeter 0–1 A; beaker of water; aerosol freezer; suitable connections and leads.

Bare the ends of the copper wire, and ensure that it can easily be connected in a series circuit with the supply and ammeter. If dry cells are used for the supply, make sure they are new ones. Provide a card giving the voltage of the supply (about 2 V will be adequate).

METHOD

1. Connect the long length of copper wire provided in series with the supply, and the ammeter. Lay the wire out in a large single loop on the bench. Read the ammeter, and record the value of the current.

Ammeter reading ... A

2. Make a rough estimate of the resistance of the wire, and explain how you do it. Suggest one reason why your estimate may be wrong, and say whether it will on that account be too high or too low.

Estimate of resistance of wire ... Ω
Explanation ...
Possible source of error ...

3. Without altering the connections, crumple the copper wire into a small tight bundle, and observe the ammeter for a minute or two. Note what happens. Then put the bundle of wire into cold water in a beaker, and again note any change in the ammeter reading. Finally, take the bundle out of the water and spray it with the freezer for a second or two, and note the effect on the reading of the ammeter. Offer explanations of your observations.

Observations:
When wire is made into a bundle ...
When wire is put in water ...
When wire is cooled by the freezer ...
Explanations ...

## 92. OSCILLATIONS OF A BAR

APPARATUS: metre rule; large G-clamp; 2 small G-clamps or 2 masses of 0·1 to 0·2 kg able to be fixed to the rule; stop watch; top-pan balance; spring balance calibrated in newtons; short rule to measure displacements; sellotape.

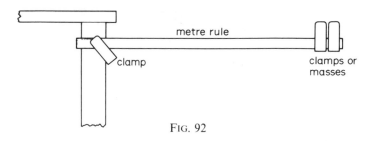

FIG. 92

Clamp the metre rule firmly near one end to a rigid table leg, so that the rule can vibrate in a horizontal plane (Fig. 92). Provide two small G-clamps (mass about 0·15 kg) which can be clamped to the rule near the freely vibrating end, or two masses of 0·1 to 0·2 kg which can easily be fixed firmly to the rule (slotted masses may serve). Provide cards giving the masses of the metre rule and of the two small clamps (or masses).

METHOD

1. The metre rule is clamped so that it can vibrate sideways. Make it vibrate, and find the time of one oscillation, using the stop-watch. It is not easy to do this, as the rule vibrates quickly. Say how you did it.

Time of one oscillation of ruler alone ...          Method used ...

2. Fix one mass (it may be a clamp) very near the free tip of the rule. Measure the time of oscillation. Fix a second mass (or clamp) as near as you can to the first, and measure the time of oscillation again. Calculate the ratio of the larger time to the smaller. A simple theory, which may not exactly apply to this case, suggests that the ratio might be $\sqrt{2}$. Discuss whether your observations are in agreement with this or not.

Time of oscillation with one mass ...          Time of oscillation with two masses ...
Ratio ...                                       Agreement or not with theory ...

3. Pull the tip of the rule to one side, using a spring balance, and observe the force needed to deflect the tip of the rule by 50 mm. Calculate the force per metre displacement, which we shall call $k$.

Force to deflect tip by 50 mm ... N          Force per metre displacement, $k = ...\ \mathrm{N\,m^{-1}}$

4. If the rule with masses at its tip were behaving like those masses restrained by springs of force constant $k$, the time $T$ of oscillation with mass $m$ at the tip would be given by

$$T = 2\pi\sqrt{m/k}$$

so that
$$m = k(T/2\pi)^2$$

Use the measured values of $k$ (from 3) and $T$ for one mass (from 2) to calculate a value of $m$, using the above equation. Look at the value of the mass provided for you. Is the calculated value too large or too small? Suggest reasons for any discrepancy.

Calculation of $m$ from $T$ and $k$ ... kg          Comparison with actual value ...
Suggested reasons ...

## 93.  STANDING WAVES ON ELASTIC

APPARATUS: Vibrator and signal generator with low-impedance output (or 'ticker timer' with vibrating reed, and power supply); 0·5 m of thin shirring elastic; curtain ring (or wire loop); 1 m stand and clamp; metre rule.

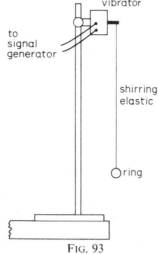

to
signal
generator

vibrator

shirring
elastic

ring

FIG. 93

Clamp the vibrator at the top of the stand, and fix one end of the shirring elastic to its vibrating arm, so that the elastic hangs beyond the edge of the bench (Fig. 93). Tie a light ring to the bottom of the elastic. Set the signal generator to about 100 Hz, and adjust the gain control until standing waves of reasonable amplitude appear when the elastic is stretched by pulling on the ring. Make sure the elastic is well lit, so that the waves can be seen.

METHOD

1. The length of elastic is vibrated at the top end. Lift the ring on the bottom end until it is slack, and then lower the ring, pulling down as necessary to stretch the elastic slowly, up to about 1 m in length. Observe closely the patterns of standing waves you see on the elastic, counting the number of 'loops' (from node to node) in each pattern.

Observations …

2. Suggest reasons why there are *more* 'loops' in the standing waves when the elastic is short and slack.

Reasons …

3. Choose two different standing wave patterns, and measure the length $l$ of the thread for each, the number $n$ of 'loops' in the pattern, and record the frequency $f$ of the vibrations. Calculate the wavelength in each case, from the fact that the length of a loop between two nodes ($l/n$) is half a wavelength. Then calculate the speed of the waves on the elastic in both cases, from $v = f\lambda$.

Length $l$ … and …
Number of loops $n$ … and …

Frequency $f$ …
Wavelength $\lambda$ … and …
Wave speed $v$ … and …

4. Do your measurements suggest that the two wave speeds are the same, or not? Give some theoretical reasons for thinking that they should be the same, or that they should be different.

Comparison of measured speeds …
Reasons for expecting speeds to be the same or not …

228

## 94. DIFFRACTION GRATING

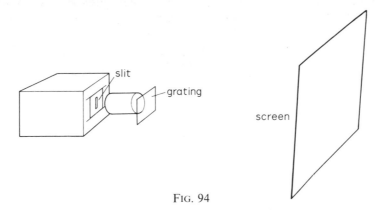

FIG. 94

APPARATUS: 35 mm slide projector; slide with slit 1 mm by 10 mm; diffraction grating (about 100 lines per mm); white screen; metre rule; red and blue filters.

Put the slide with a slit in it (a slit cut in a square of card will do) in the projector, and focus the image of the slit on the screen, with the screen about 1 m from the projector. Mount the grating (if necessary) so that when it is held just in front of the projector lens, no light goes round the outside of the grating. Darken the area where the experiment is set up.

METHOD

1. The slide projector should cast a sharp image of a slit on the screen; adjust the focus if necessary. Then hold the grating just in front of the projector lens, so that all the light goes through the grating. Turn the grating round, and by observing the pattern on the screen, find the position in which the rulings of the grating are parallel to the slit. Describe what you see as you turn round the grating, and explain how you choose the position when slit and rulings are parallel.

    Observations ...
    Explanation ...

2. Put a red and then a blue filter in front of the grating, and describe what happens to the pattern on the screen. Explain the changes you see.

    Observations ...
    Explanation ...

3. Measure the distances, on either side, from the white central image to (i) the first red image, and (ii) the first blue image. Work out the ratio of the difference between these distances to their average distance. Explain why this is an estimate of the ratio of the change in wavelength between red and blue, to the average of the two wavelengths.

    Distance to red image ...
    Distance to blue image ...
    Ratio of difference in distances to average distance ...
    Explanation of relationship to wavelengths ...

## 95.  LIGHT AND A PHOTOCELL

APPARATUS: Motor headlamp bulb, 12 V, 24 W; power supply variable from 6 V to 15 V; ammeter 0–5 A; voltmeter 0–15 V; photocell and suitable microammeter or milliammeter; black screens.

The experiment requires a dark room. Connect the lamp to the supply, with ammeter and voltmeter. Mount the photocell at such a distance from the lamp that the meter connected to the cell indicates nearly full-scale when the lamp is overrun at about 15 V. Different types of photocell are suitable, including photoresistive cells or phototransistors, both of which need a suitable source of current.

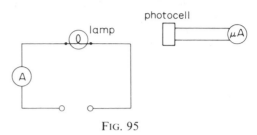

FIG. 95

METHOD

1. Check that, with the supply for the lamp at 12 V, the meter connected to the photocell gives an indication. Turn off the lamp and see if there is any reading on the meter when the photocell is in darkness. If there is, subtract this reading from future readings. Turn on the lamp, and arrange black screens so that the light reaching the cell is not affected by your own movements (as you bend to look at meters, for example).

Move the lamp towards and away from the photocell, and explain what you observe.

Observations ...                    Explanation ...


2. Fix the lamp roughly in its original position and read the photocell current, the current in the lamp, and the voltage across the lamp (about 12 V). Repeat these measurements for a lower voltage (perhaps about 9 V) at which the lamp is quite dim; and for a higher voltage (up to 15 V) at which it is very bright. (Do NOT leave the lamp running at 15 V for long, or it will burn out.)

|  | Normal voltage | Low voltage | High voltage |
|---|---|---|---|
| Current from photocell | ... mA or $\mu$A | ... mA or $\mu$A | ... mA or $\mu$A |
| Current in lamp | ... A | ... A | ... A |
| Voltage across lamp | ... V | ... V | ... V |

3. Calculate the power delivered to the lamp when the voltage is lower than normal, and when it is higher than normal. Work out the ratio of the two powers. Work out the ratio of the currents from the photocell in the two cases. Explain whether your results support the view that the two ratios are the same, or not.

Power to lamp ... W and ... W              Ratio of currents from photocell ...
Ratio of powers ...                        Comparison of the two ratios ...

4. The two ratios might not be equal for several reasons. One set of reasons could have to do with the amount of light given out by the lamp at various voltages (at low voltages it produces only heat). Another set of reasons could have to do with the way the photocell responds to different intensities of light. Explain in greater detail your suggestions for reasons of these two kinds, and indicate which reasons you think likely to hold good in the present case.

Reasons to do with the lamp ...            Reasons to do with the photocell ...

## 96. HEATING AND COOLING

APPARATUS: Aluminium block A of mass 1 kg (e.g. Nuffield Physics item 77) and aluminium block B of mass about 100 g, as nearly as possible the same shape and proportions as the larger block; 2 thermometers 0–100 °C; large pan of hot water; expanded polystyrene or cork mat; ruler; top-pan balance; two stop-clocks. Both blocks must be drilled to take a thermometer.

Keep the two blocks in hot water (over a low flame or with an immersion heater if necessary) for about 10 minutes before the experiment begins. The water need not be boiling.

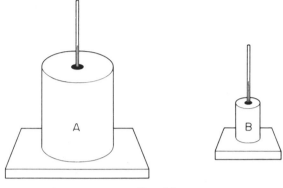

FIG. 96

METHOD

1. Check that the two blocks in the hot water are at roughly the same temperature by putting thermometers in the holes in each block. Then take the blocks out of the water, and stand them (not too close) on insulating mats.

Use the two stop-clocks to record the time taken for each block to cool from one temperature to another about 5 °C below it. Make the starting temperature as high as possible, to save time:

|  | Large block | Small block |
|---|---|---|
| Time to cool from ... °C to ... °C | ... s | ... s |

2. Weigh the two blocks, and work out the ratio of their masses. Measure their dimensions, and work out the ratio of their *surface areas*. Explain how you work out the ratio of the areas. Explain why the two ratios are not the same, even though the blocks are made of the same material (aluminium).

Masses ... kg and ... kg          Ratio of surface areas ...
Ratio of masses ...               Explanation of calculation ...
Dimensions ...                    Reason why the two ratios are different ...

3. Suppose that the amount of energy radiated and convected away per second is proportional to the surface area, at a definite temperature. If so, what is the ratio of the energy (heat) lost per second by the two blocks? Which loses the larger amount of energy per second?

Ratio of energy lost per second ...          Explanation ...

4. If the two blocks lost the same amount of energy, which would cool by the greater amount? Explain. Use this explanation to suggest how to explain the ratio of the two rates of cooling observed in 1. If possible, calculate the ratio from the data in 2 and 3, and compare it with what you observed.

Explanation of cooling when the same energy is lost ...
Explanation of why one block cools more quickly than the other ...

231

## 97. FORCES, ACCELERATIONS AND ENERGY

APPARATUS: Steel ball-bearing (about 15–20 mm diameter); plastic curtain track; metre rule; top-pan balance.

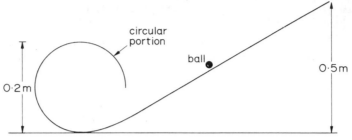

FIG. 97a

The track must be *firmly* fixed in the shape shown in Fig. 97a, perhaps by being glued or screwed to projections from a vertical back-board. Fixing points should not be more than 0·2 m apart, preferably less. Screws must be countersunk so as not to affect the motion of the ball on the track.

METHOD

1. Roll the ball down the track, starting at greater and greater heights. Describe what happens, and say what you can about it in terms of changes of energy, and of forces acting on an object moving round in a circle.

Description ...                    Further explanation ...

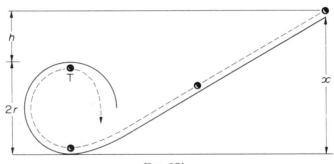

FIG. 97b

2. Find that height $x$ of the ball above the bench when it rolls to the top T of the circular part of the track, and then *just* leaves the track at T, and falls freely through the air (Fig. 97b). Measure the diameter $2r$ of the circular path of the ball, as shown. What is their difference, $h$?

Height $x$ ... m                    Diameter $2r$ ... m                    Difference $h$ ... m

3. When the ball just leaves the track at the top, the downward force of gravity on it, $mg$, is just large enough to provide the inward force needed to keep it moving in a circle of radius $r$, which is $mv^2/r$. Use this to work out the speed of the ball at the top of the track. Weigh the ball, and calculate its kinetic energy (neglecting rotation), at that point.

Speed of the ball at the top ... $m\,s^{-1}$          Kinetic energy ... J

4. Describe the energy changes as the ball rolls along the track, and explain why the kinetic energy you have just found cannot be more than $mgh$, where $h$ is the difference between $x$ and $2r$, found in 2.

Description ...          Explanation ...

5. Calculate the energy change $mgh$ and compare it with the kinetic energy obtained in 3. Suggest reasons for any difference, and ways of testing whether those reasons are correct.

Energy change $mgh$ ... J          Reasons for difference ...
Comparison with kinetic energy ...          Tests of reasons ...

## 98. STRETCHING RUBBER BANDS

APPARATUS: assorted rubber bands (bought in the same packet), metre rule; S-hooks; G-clamps; spring balances 0–10 N.

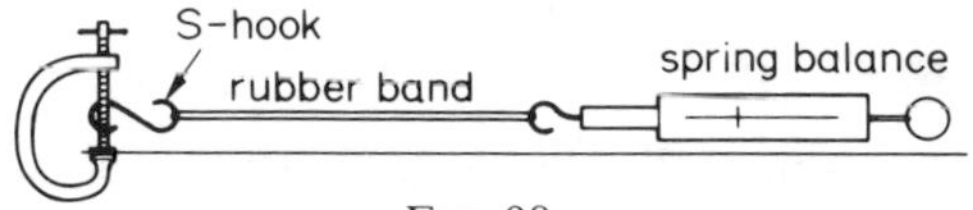

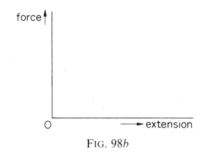

FIG. 98a

Fix one S-hook firmly, for example with a G-clamp, so that rubber bands can be attached to it and pulled horizontally by a spring balance (Fig. 98a).

METHOD

1. Stretch a rubber band by pulling on it with the spring balance, and continue until it is difficult to stretch it any further. Without taking more than a few rough readings, make a rough sketch of the shape of the graph of force against extension as the band is stretched (Fig. 98b). Explain the shape you have drawn.

FIG. 98b

Explanation ...

2. Measure the extension of the rubber band at some force large enough to roughly double the length of the rubber. Now choose another band which has the *same* cross-section, but is longer or shorter than the first. Measure the extension at the *same* force as before. Explain as fully as you can any difference in the two extensions.

Force ... N                                  Original lengths ... m and ... m
Extension ... m and ... m                    Explanation of any difference in extension ...

3. Choose two bands of the same length, but different thickness or width. Measure the two thicknesses and widths, and work out the cross-sectional areas of the two bands. Stretch each to the *same* extension (about doubled in length) and note the force needed in each case. Explain as fully as you can any difference in the two forces.

Extension ... m                              Forces ... N and ... N
Cross-sections ... mm$^2$ and ... mm$^2$     Explanation of any difference in force ...

234

## 99. ROTATIONAL MOTION

APPARATUS: horizontal turntable (gramophone turntable or wooden disc mounted on a cycle wheel, for example); two 1 kg masses; stop-clock; device to produce audible sound about once a second, e.g. metronome or electric stop-clock; metre rule.

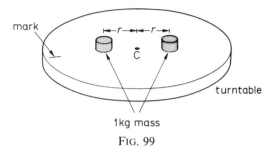

FIG. 99

Fix the turntable firmly so that it is free to turn in a horizontal plane, and choose masses which rest stably on the turntable.

METHOD

1. Put the two masses on the table, close to the centre C and exactly opposite each other at the same distance $r$ from C (Fig. 99). Practice turning the turntable with a finger so that it goes round exactly once for every tick of the metronome or other device provided (about once per second). Then let the table turn on its own, starting the stop-clock as you do so, and measure how long it takes to come to rest. Record the distance $r$ between the centre of each mass and the centre of the turntable.

    Distance $r$ ... m                                            Time to come to rest ... s

2. Now move each mass to the *edge* of the turntable at the same distance $r$ from C, noting the new distance $r$. Repeat the previous measurement, being as careful as you can to start the turntable at the same speed. Suggest an explanation of any difference in the two times.

    Distance $r$ ... m
    Time to come to rest ... s                             Explanation of difference in times ...

## 100.  ISOTHERMAL COMPRESSION

APPARATUS: Lightweight aluminium cycle pump; aluminium block (1 kg) or other large metal block as heat sink; 36 s.w.g. copper–constantan thermocouple; light-beam galvanometer (e.g. Scalamp).

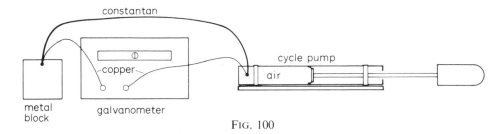

FIG. 100

Fix the cycle pump firmly to a base board with (for example) Terry clips, and clamp the base board to the bench, so that the pump can be operated without touching the casing (Fig. 100). Seal the outlet of the pump with a screw. Sellotape one junction of a copper–constantan thermocouple to the casing of the pump, and the other junction to a large block of metal (to keep it at constant temperature). Connect the thermocouple to the galvanometer, and check that the galvanometer deflects by a few mm, during the few seconds after the pump is pushed smartly in. Grease the pump washer. Put a card by the apparatus giving the mass of the pump casing, the specific heat capacity of aluminium, and the internal volume of the pump casing.

METHOD

1. Gently touch the thermocouple junction attached to the pump with one finger, to warm it, and observe the deflection of the galvanometer connected to the thermocouple. Consider what order of magnitude of temperature rise would cause a deflection of, say, 10 mm. Do not touch the casing of the pump again.

   Temperature rise for about 10 mm deflection …

2. When the galvanometer is steady again, pull out the pump handle, and then push it in steadily, so as roughly to halve the volume of air in the pump. Record the change in deflection of the galvanometer. Repeat, checking that the change in deflection is much the same.

   Deflection when air in pump is compressed …

3. Because the temperature rise of the pump is very small, it may reasonably be claimed that the compression is nearly, though not exactly, isothermal.
   If the air in the pump is initially at atmospheric pressure (about $10^5 \, \mathrm{N\,m^{-2}}$) and the volume is halved isothermally, the pressure rises to about twice atmospheric pressure. Use this information, and the volume of the casing (given on a card with the apparatus) to estimate roughly the energy transformed when the air was compressed. Explain how you make the estimate. (You can check your estimate by guessing roughly the average force exerted and the distance moved by the pump, so getting a second estimate of the energy transformed.)

   Estimate the energy transformed from pressures and change in volume …

4. Use the mass of the pump casing, and the specific heat capacity of aluminium, to estimate the temperature rise of the casing, if all the energy in 3 is given to the casing and warms it. Suggest reasons why this may not exactly be so.

   Estimate of temperature rise …
   Reasons why not all the energy goes to the casing …

# Miscellaneous Examination

# Questions and Investigations

# Miscellaneous Examination Questions

[Two experiments may be performed in 3 hours.
The *Apparatus* for each experiment is given at the end.]

## MECHANICS AND PROPERTIES OF MATTER

**1.**

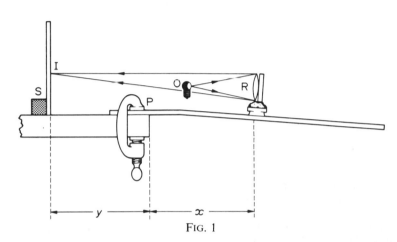

FIG. 1

*1.* Clamp a metre rule to the bench (or other suitable mounting) so that 800 mm of the rule overhangs the edge of the bench. Denote the position of the edge of the bench on the metre rule by the point P.

*2.* R is a reflector system, tripod mounted, ready for your use. Place R on the clamped rule in a position such that the front surface of the lens is a distance $x$ from P where $x$ is between 300 mm and 700 mm.

*3.* Place the image screen S on the bench at the distance $y$ from P so that it faces R along the length of the clamped rule. (The value of $y$ will be given.)

*4.* Mount the luminous object O in a suitable stand.

*5.* Position O between S and R, keeping O within 50 mm of the top surface of the rule, so that a sharp image I of O is formed by R on S. (Gently slide the card C between the lens and the mirror to check that you have observed the correct image; if correct, the card will make the image vanish. Raise the card before proceeding further.)

*6.* Record the position of I on S.

*7.* Place the 100 g mass $M$ on the clamped rule so that its centre is 20 mm from the free end.

*8.* Record the new position of I on S.

*Note:* In order that the images in (6) and (8) shall lie on the screen it may be necessary to tilt the lens and mirror in the plasticine mounting. This action, when necessary, must precede the sequence of operations (6) to (8), otherwise the tilting will invalidate the value of Z obtained in (9) below.

*9.* From (6) and (8) obtain the distance Z between the two positions of the image I, taking Z to be positive if the image in (8) lies above the image in (6), and negative if it lies below.

*10.* Record the value of $x$.

*11.* Remove the mass $M$.

*12.* Move R to a new position along the rule and repeat instructions (5) to (11) taking care to keep S at the distance $y$ from P.

*13.* Repeat instruction (12) to obtain a series of corresponding pairs of values for Z and $x$, where $x$ lies between 300 mm and 700 mm.

*14.* Remove R from the rule.

*15.* Measure D, the change in position of the free end of the rule against a suitably placed vertical scale when the mass $M$ is placed at this end as in (7).

*16.* Plot a graph with Z as ordinate and $x$ as abscissa.

*17.* Record the slope G of the graph at the point where Z is zero.

*18.* Calculate G/D.

(*Apparatus:* Illuminated crosswires O. Metre rule. G-clamp. Mass M 100 g. Screen S with surface covered with metric graph paper. Card with value of distance SP or $y$ to form real image on S as shown. Card C. Suitable converging lens with plane strip mirror behind for R, in plasticene mounting on tripod.)

**2.** Measure the vertical displacement $h$ of the bob of

239

a simple pendulum when it is displaced through a horizontal distance $d$. Plot a graph of $d^2/d$ against $h$ and use it to find the length of the pendulum. Check your answer by determining the period of oscillation $T$ of the pendulum. (You will be given the value for the gravitational acceleration.)

(*Apparatus:* Simple pendulum attached to 'inaccessible' point. Thread, clamp, and stand for holding bob in displaced position. Metre rule clamped horizontally immediately below bob. Set square and half-metre rule for vertical measurement. Stopwatch or clock.)

**3.** A metre rule is suspended horizontally by two vertical threads of equal length. Keeping the threads vertical, arrange them symmetrically with respect to the centre of the rule at a measured distance $a$ apart. Find the time period $T_1$ for small oscillations of the rule in the vertical plane in which it lies when the system is at rest. Also find the time period $T_2$ of small rotational oscillations of the rule about a vertical axis through its centre. Make further determinations of $T_2$ for different values of $a$ and calculate $T_1/T_2$ for each case. Plot a graph of $T_1/T_2$ as ordinate against $a$ as abscissa and record its slope.

(*Apparatus:* Metre rule suspended with shortest dimension vertical, the vertical threads being about 0·5 m long. Second metre rule. Pointer for defining complete oscillations.)

## HEAT

**4.** Assuming the specific latent heat of fusion of ice to be $336 \text{ kJ kg}^{-1}$, find the specific heat capacity of the given liquid.

(*Apparatus:* Shielded calorimeter and stirrer. Balance and weights. Thermometer 0–50 °C in $\frac{1}{10}$° or $\frac{1}{5}$ °C. Ice. Blotting or filter paper for drying ice. Stop-watch or clock. Beaker with light oil, e.g. Castrolite. Tripod, gauze, burner. Card stating thermal capacity of calorimeter and stirrer in $\text{J K}^{-1}$.)

**5.** Observe the fall of temperature which occurs when 25 g of 'hypo' are dissolved in $100 \text{ cm}^3$ of water contained in a calorimeter, all initially being as nearly as possible at air temperature. Then use the block of metal of known heat capacity to determine the combined heat capacity of the solution and calorimeter. Hence calculate the quantity of heat absorbed when the 'hypo' was dissolved.

(*Apparatus:* Shielded calorimeter. Beaker with water at air temperature. Measuring cylinder. Balance and weights. Watch-glass and filter paper for weighing and handling 'hypo'. Supply of 'hypo'. Metal block (e.g. 200 g), with attached thread. Beaker, tripod, gauze, burner. Glass rod for stirring solution. Thermometer 0–100 °C, thermometer 0–50 °C in $\frac{1}{10}$° or $\frac{1}{5}$ °C for calorimeter ONLY. Card stating heat capacity of metal block in $\text{J K}^{-1}$.)

## OPTICS

**6.** By forming a diverging liquid lens below a converging lens for each liquid, estimate the refractive index of glycol assuming that for water is 4/3.

(*Apparatus:* Converging lens 10–15 cm focal length, thin plane mirror, pin, stand and clamp, metre rule, small pipette, water, glycol.)

**7.** Use the converging lens to form a magnified real image of the illuminated object on the screen, and measure an appropriate linear dimension ($y_1$) of the image. Keeping the object and screen fixed, shift the lens so as to form a diminished image and measure the same dimension ($y_2$). Repeat these observations for several different separations of the object and screen. Plot a graph of $y_1$ against $1/y_2$ and find it slope.

(*Apparatus:* Biconvex lens about 20 cm focal length, in holder. Illuminated object such as perspex ruler, or two thin parallel wires about 1·5 cm apart. White screen. Metre rule. Pair of dividers.)

**8.** Put some water in a cylindrical vessel and allow a small pin $P$ to rest on the bottom. Place a strip of plane mirror across the top of the vessel and adjust the position of another pin $Q$ until the image of $Q$ in the mirror coincides with the apparent position of $P$ when viewed from a position vertically above $P$. Repeat this for various depths $h$ of the water. Plot $h$ as ordinate against the distance of $Q$ above the mirror as abscissa. Find the slope of the graph.

(*Apparatus:* Wide tall gas jar. Piece of plane mirror to rest across part of top of jar. Small pin. Large pin. Clamp and stand. Metre rule. If necessary, low stool for resting gas jar.)

## SOUND

**9.** Using the resonance tube, find the first position of resonance for each of the given tuning forks. Plot a graph of the resonant length against the reciprocal of the frequency, and use the graph to determine the velocity of sound in air at room temperature. Record the value of the intercept on the axis of length.

(*Apparatus:* Resonance tube closed at one end, means of altering resonant length. Set of at least 5 tuning forks with known frequencies. Striking pad, metre rule.)

**10.** Keeping the tension on the sonometer wire A constant find the length $l$ of A which vibrates in unison with a fixed length of wire B for various values of the tension $T$ in B. Assuming the relation $l = kT^n$ determine by a graphical method the value of $n$.

(*Apparatus:* Sonometer with usual accessories, one wire B for varying tension, comparison wire A, five known weights.)

## ELECTRICITY

**11.** Observe the terminal p.d. of a dry cell at 1-minute intervals immediately after connecting the cell to a low series resistor R.

Calculate the internal resistance $r$ of the cell with increasing time and discuss the variation of $r$.

(*Apparatus:* Potentiometer and accessories as in Experiment 62 to measure internal resistance; stopclock.)

**12.** Using a simple wire potentiometer, determine the resistance of the given wire P by comparison with a standard coil, and hence the resistivity of the material of the wire. Repeat the experiment using a different length of the same wire and a different standard resistance.

(*Apparatus:* Potentiometer and accessories. P to be about 60 cm s.w.g. 26 constantan, other length about 150 cm of same wire. $2\,\Omega$ and $5\,\Omega$ standard coils, micrometer gauge, metre rule, another accumulator, two rheostats, suitable terminals.)

**13.** Compare the diameters of the wires X and Y, which are made of the same alloy, by determining the resistances of equal measured lengths of the wires with a metre bridge. The result should be obtained from a graph of the resistance of X plotted against the resistance of an equal length of Y. Check your answer by means of a screw gauge.

(*Apparatus:* Metre bridge and accessories, metre rule, micrometer gauge, standard $1\,\Omega$ coil. X and Y to be about 1 metre of No. 24 and No. 28 constantan respectively.)

**14** Fig. 2(a) shows that part of the apparatus which is provided already assembled. Set up the circuit shown in Fig. 2(b) and connect the leads A and B into the circuit as shown.

Starting with Q set at 90 ohms, reduce Q until the smallest deflection in the galvanometer G is obtained with the deflection in the same direction as at first. Raise the temperature of the contents of the beaker until G shows no deflection. Take the precautions you

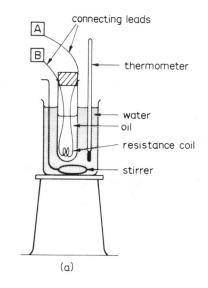

(a)

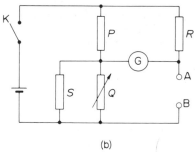

(b)

Fig. 2

think are necessary to ensure uniformity of temperature throughout the immersed resistance coil.

Record the value of $Q$ and the temperature $\theta$ registered by the thermometer. Increase the value of $Q$ to obtain a set of corresponding values of $Q$ and $\theta$.

Plot a graph with $Q$ as ordinate and $\theta$ as abscissa.

Determine the slope $z$ of the curve at a convenient point where the value of $Q$ is $Q_0$.

Calculate $z/(S + Q_0)^2$.

(*Apparatus:* As in Fig. 2. Resistance coil of fine iron or other wire of a few ohms. $R$ of similar resistance to coil. Suitable variable resistance $Q$ in steps of 1 ohm to 100 ohms. $P$ of order 10 ohms. $S$ of order 20 ohms. Suitable centre-zero galvanometer G with protective resistance.)

**15.** By means of a potentiometer measure the difference between the e.m.f. of a lead accumulator and that of a cell X of known e.m.f. [Only one accumulator must be used to supply current to the potentiometer wire.]

Also, obtain a value for the internal resistance of the cell X.

(*Apparatus:* Potentiometer and usual accessories. Another lead accumulator, X is dry (or other type) cell e.m.f. stated on card, known resistances of 1, 2, 2, 5 Ω. Connecting wire. Two plug keys.)

**16.** Use a metre bridge to determine the temperature coefficient of resistance of the wire in the given coil.

A thermometer is not provided, but it may be assumed that melting ice and boiling water in the laboratory define the lower and upper fixed points of a Centigrade scale. Determine room temperature on this scale.

(*Apparatus:* Metre bridge and accessories, 2 Ω comparison coil. Crushed ice, beaker with water. Coil of enamelled copper wire, or nickel or 'black' iron wire in test-tube with oil, of resistance between 1·5 and 2·5 Ω, means for boiling water.)

**17.** Connect a thermocouple, a galvanometer and a resistance box in series. Adjust the resistance of the box so that the galvanometer shows a convenient deflexion when one junction of thermocouple is in melting ice and the other is in water boiling at atmospheric pressure. Observe this deflexion and then use the apparatus to determine the boiling-point of brine on a Centigrade scale based on the thermocouple. [A change of 1 mm of mercury in barometric pressure alters the boiling-point of water by 0·037 °C.]

(*Apparatus:* Copper–constantan thermocouple. Pointer galvanometer and resistance box. Melting ice. Means for boiling water. Access to barometer. Saturated brine.)

## ADDITIONAL

*1.* Place the piece of wood or chipboard on the bench to provide a flat surface on which to bounce the table tennis ball. (See *Apparatus*.)

*2.* Using the clamp and stand, mount the metre rule vertically so that heights above the top surface of the wood can be measured.

*3.* Release the ball from rest so that it falls through a distance $l_1 = 40$ cm to the surface of the wood.

*4.* Determine the distance $l_2$ through which the ball rises after bouncing once on the wood.

*5.* Repeat steps *3* and *4*, for different values of $l_1$, to obtain a series of corresponding values of $l_1$ and $l_2$.

*6.* Plot a graph of $l_2/l_1$ as ordinate against $l_1$ as abscissa.

*7.* Use your graph to obtain a value for $l_2/l_1$ when $l_1 = 0$.

*8.* Use your graph to predict the height reached by the ball after being released from rest at $l_1 = 60$ cm and making *two* consecutive bounces.

*9.* Check your prediction in *8* experimentally, recording all measurements made.

(*Apparatus:* Piece of smooth wood or chipboard, thickness 17 mm or more and area approximately 200 mm × 200 mm. Table tennis ball (Halex 2-star or equivalent). Metre rule. Clamp and stand to support the rule in a vertical position. Additional boss, on the same stand, supporting a horizontal rod 100 mm to 150 mm in length.)

# Suggestions for Investigations

The following suggestions are for subjects or problems you might investigate. Each is a starting point, rather than a goal you should expect to reach; many could lead to several different investigations connected with various aspects of the problem.

Choose a problem which interests you, try out some rough ideas quickly, and then settle on one definite aspect to investigate more thoroughly. It will often turn out that some of your first ideas are good, others are bad, and some are good but too ambitious. Very often, simple methods are no worse than sophisticated ones, and improvised apparatus as good for the purpose as an expensive ready-made item.

There are some rough and ready rules worth bearing in mind:

(a) Work out what apparatus and materials you think you want, so as to make sure they are obtainable.

(b) Make a rough plan of what you are going to do in the first stages.

(c) It is best to get on quickly and try out something in practice; one may learn more that way than by making very elaborate plans. Make some measurements as soon as you can, so that you have something definite to think about.

(d) When something unexpected happens, don't ignore it or suppose that 'it must be wrong'. Following up surprises often leads to new discoveries.

## 1. *Windy pedestrian places*

It is found that an open pedestrian space between a tall tower block A and a lower one B is much windier than the architects expected, when the wind blows on to the blocks as in Fig. 1. Investigate why this happens, and how it depends on the size and placing of the blocks. Use models in an air stream in the laboratory, produced by say a vacuum cleaner, perhaps with paper streamers to indicate the wind direction.

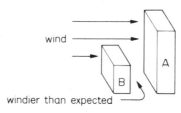

FIG. 1

## 2. *Diffraction effects*

A typewriter can be used to make regular patterns or dots on paper, and such patterns can be reduced in scale by photographing them. Investigate the diffraction effects of 'dot diffraction gratings' having various dot patterns and spacings.

## 3. *Flame speed*

The speed at which a flame travels along a strip of paper (or other material) depends on the thickness of the paper. Investigate the effect. Can the flame speed be altered by conducting heat from the paper?

## 4. *Design of arrow*

An arrow released from a bow may take different paths in air depending on the design of the tip or tail. Investigate the effects of different designs. What sort of arrow always points along its flight path rather than tumbling over and over in flight?

243

### 5. *Hall effect* (other than with metal foil)

By using very thin layers and very powerful magnetic fields, investigate if the Hall effect can be obtained with conducting paint or colloidal graphite (Aquadag) or with solutions of electrolytes.

### 6. *Insulation of hot water tank*

By using a model, investigate whether the lagging round a cylinder full of hot water is best placed uniformly over the whole cylinder or whether there is a better way of distributing a given amount of lagging.

### 7. *Lift of model aircraft wing*

Investigate the forces on a model aircraft wing when it is placed in an air stream. You may find it helpful to produce the air stream with the aid of a vacuum cleaner. Alternatively, you may find it better to use a water stream in place of an air stream.

### 8. *Short pulses of magnetic field*

In some experiments, a large but very brief pulse of magnetic field is wanted. Investigate ways of making such pulses—they should last for as short a time as possible and be large during that time. (Take care! A short sharp change of flux in a coil will induce a high voltage across the coil.)

### 9. *Stiffness of bridge girder*

Fig. 2 shows the general shape of a bridge girder. Make some balsa wood models, and investigate how their bending under load varies with such factors as the number, placing, angle and thickness of the struts.

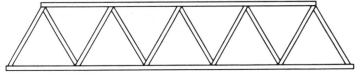

FIG. 2

### 10. *Recorder as simple tuned pipe*

Investigate how the note obtained from a recorder varies with the length of the open air column. Does it behave like the simple pipe described in textbooks? How, if at all, does the frequency of the note vary with the strength with which the instrument is blown?

### 11. *Smoke clinging to glass*

A smoky taper held very close to a vertical pane of glass for a short time, and then removed, will often leave behind a layer of smoke close to the glass which persists for a time, sometimes creeping up the glass. Investigate the cause of the effect and the factors which influence the time the layer persists.

### 12. *Ultraviolet light through glass.*

Short-wavelength ultraviolet light is strongly absorbed by glass but some radiation, too short in wavelength to be visible, does penetrate the glass. Investigate at what wavelength the glass stops the ultraviolet light, and whether the cut-off point is altered on varying the thickness of the glass. You may find it convenient to detect the radiation photographically.